A Handbook for
Stream Enhancement & Stewardship

The Izaak Walton League of America

For more than thirty years, the Izaak Walton League's watershed stewardship programs have developed innovative educational programs for groups and individuals. The League has educated and motivated citizens to clean-up stream corridors, monitor stream health, restore degraded stream banks, and protect dwindling wetland acreage.

These important watershed stewardship activities have been implemented in communities across the country through the League's more than three hundred local chapters. For more information about the League's watershed stewardship programs, resources, and technical assistance, please call the program hotline at **1-800-BUG-IWLA (284-4952)**, send an email to **sos@iwla.org**, or visit the web site at **www.iwla.org/sos.**

Founded in 1922, The Izaak Walton League of America is dedicated to common-sense conservation that protects America's hunting, fishing, and outdoor heritage relying on solution-oriented conservation, education, and the promotion of outdoor recreation for the benefit of our citizens. The League has more than 40,000 members and supporters in 21 state divisions and more than 300 local chapters in 32 states.

The Izaak Walton League of America
707 Conservation Lane
Gaithersburg, MD 20878-2983
(800) BUG-IWLA
sos@iwla.org
www.iwla.org/sos

A Handbook for

Stream Enhancement & Stewardship

by

The Izaak Walton League of America

A joint publication of

The McDonald & Woodward Publishing Company
Blacksburg, Virginia

and

The Izaak Walton League of America
Gaithersburg, Maryland

A Handbook for Stream Enhancement & Stewardship

A joint publication of

The McDonald & Woodward Publishing Company
Blacksburg, Virginia, and Granville, Ohio
www.mwpubco.com
            and
The Izaak Walton League of America
Gaithersburg, Maryland
www.iwla.org

Printed in The United States of America by McNaughton & Gunn, Inc.

The first edition of this book was published as *A Citizen's Streambank Restoration Handbook* in January, 1995. A supplement to that handbook, *Restoring the Range: A Guide to Restoring, Protecting and Managing Grazed Riparian Areas,* was published in May, 1995.

Second Edition, April 2006.

15 14 13 12 11 10 09 08 07 06   10 9 8 7 6 5 4 3 2 1

Design based on concepts by DiGiorgio Designs and layout on concepts by Pam Cullen.

This book was produced with a generous grant from the Norcross Wildlife Foundation.

Library of Congress Cataloging-in-Publication Data

A handbook for stream enhancement & stewardship / The Izaak Walton League of America.— 2nd ed.
    p. cm.
  "The first edition of this book was published as A Citizen's Streambank Restoration Handbook in January, 1995"—T.p. verso.
  Includes bibliographical references.
  ISBN 0-939923-98-X (alk. paper)
  1. Stream restoration. 2. Riparian areas. I. Title: Handbook for stream enhancement and stewardship. II. Izaak Walton League of America.
  QH75.H364 2006
  627'.12—dc22

                                                                2006004309

Table of Contents

Acknowledgments

The first edition of this handbook was written by Karen Firehock and Jacqueline Doherty, and a supplement that followed was written by Jay West. The present edition was written by Leah Miller, Gwyn Rowland, and Casey Williams, and Suzanne Zanelli assisted with various parts of the text and illustrations. Expert advice on various portions of the text of the present edition was provided by the following reviewers:

• Carolyn Adams, Director, USDA-NRCS Watershed Science Institute, Raleigh, NC

• Jerry Bernard, National Geologist, USDA-NRCS, Washington, DC

• Ann Butler, Water Quality Program, WA Department of Ecology, Olympia, WA

• John A. Conners, Consulting Geologist, South Otselic, NY

• Drew DeBerry, American Forest Foundation, Washington, DC

• Howard Hankin, National Aquatic Ecologist, USDA-NRCS, Washington, DC

• Tim Hoffman, Environmental Quality Resources, Inc, Gaithersburg, MD

• Chad Moore, Physical Scientist, Pinnacles National Monument, National Park Service, Paicines, CA

• John S. Moore, National Fluvial Geomorphologist/Hydrogeologist, USDA-NRCS, Washington, DC

• Tom Noonan, Watershed Planner, USDA-NRCS Watershed Science Institute, Raleigh, NC

• Dennis O'Connor, Restoration Ecologist, Portland, OR

• Annie Phillips, Watch over Washington, WA Department of Ecology, Olympia, WA

• John Potyondy, Program Manager, Stream Systems Technology Center, Fort Collins, CO

• Kerry Robinson, Hydraulic Engineer, USDA-NRCS Watershed Science Institute, Raleigh, NC

• Barry Rosen, Water Quality Specialist, USDA-NRCS Watershed Science Institute, Raleigh, NC

• Rob Sampson, State Conservation Engineer, USDA-NRCS, Anchorage, AK

• Jon Werner, National Hydraulic Engineer, USDA-NRCS, Washington, DC

Introduction

A **stream** is a body of running water flowing through a clearly defined natural **channel** to progressively lower levels. **Rivers**, creeks, brooks, and runs are all streams. Streams also possess environmental, social, cultural, and economic value. They provide humans with water for drinking, irrigation, industry, power production, transportation, flood control, fishing, boating, swimming, and aesthetic enjoyment.

Streams are dynamic systems; they function and evolve in response to what is happening in other parts of their ecosystem. Changes within the surrounding ecosystem (i.e., other parts of the **watershed**) affect the physical, chemical, and biological processes occurring within a stream. Human communities can be instrumental in protecting and enhancing stream **habitat** and water quality. Citizens can play a critical role in educating planners, engineers, businesses, and others about stream behavior, and they can propose new ways to manage our waters. This handbook is intended to help citizens, communities, organizations, companies, and governments understand, recognize, and take an active role in planning for environmentally sound, cost-effective stream **enhancement**. Taking action now will help to ensure that future generations enjoy the beauty, recreational opportunities, drinking water supply, and other benefits of our watersheds.

Streams are unique features that connect landscapes and communities, providing unlimited opportunities to bring people together to create a common vision for productive and sustainable conditions. Successful management of streams is dependant on bringing communities of people together, including public and private stakeholders and government staff, to coordinate resources and design effective solutions (Figure 1).

> Words presented in bold type are defined in the Glossary on pages 113–119.

Figure 1. Volunteers can participate in many aspects of stream enhancement projects. These community members are contributing to a project by monitoring site conditions. (Izaak Walton League of America photograph)

This handbook can help citizens and communities guide streams back to a healthy and balanced state within current watershed conditions. The handbook focuses on ecological enhancement because **restoration** may not be possible or desirable. True restoration means returning an ecosystem to a close approximation of its condition prior to disturbance. Landscape changes in the watershed may no longer support previous conditions. It is often impossible to achieve ecological restoration, especially in areas where land uses and infrastructure such as roads, buildings, and water-control structures are established. Nevertheless, stream condition can be enhanced through **structural** and **non-structural** work. Structural techniques involve recreating the shape of the bank and often include adding materials such as rock to harden the **stream bank**. Non-structural designs include incorporating conservation measures, such as strategic grazing, or planting riparian vegetation to minimize the effects of current land use (Figure 2).

Every stream has unique characteristics and requires a unique enhancement strategy. Local natural resource professionals can provide insight into stream conditions in a community. Community members can play a key role in planning stewardship projects and gathering the information needed to design enhancement projects.

The following chapters provide citizens with a crash course in the science behind stream systems, the techniques used to assess a watershed and inventory the

Figure 2. Stable stream systems rely on complex interactions between their living and non-living components. Native riparian vegetation and rocks along the bank protect this stream corridor from degradation. (Bill Weihbrecht photograph)

health of a stream, and the basic principles and planning of stream enhancement. They also explore ways to enhance stream banks and grazed rangeland, including the design of improvement techniques. This book is not a comprehensive manual for professionals trained in **stream restoration**, but rather a resource that volunteers can use to become informed participants and organizers of enhancement programs. The Izaak Walton League recognizes that any attempt to present a complex issue in a manner that is instructive and easy to understand will result in the generalization of some topics. Please consult the Bibliography at the end of the book and the Watershed Stewardship Resources document available on-line at *www.iwla.org/sos/resources* for more information.

Chapter One

Overview of Stream Systems

*T*his chapter describes how water flows into and through different parts of stream systems. Understanding how fast, how much, how deep, how often, and when water flows is important when considering a stream enhancement project. This chapter summarizes a vast amount of background information needed when discussing stream health and introduces technical concepts and terms that describe stream behavior. Note that the words stream and channel are used interchangeably throughout this publication. Warning: This technical language can be dense and dry, but it's worth learning. Technical information provides a foundation of knowledge used to investigate stream problems and it will assist a group when writing grants and communicating with natural resource professionals.

Streams

As mentioned in the Introduction, *stream* is an umbrella term used by scientists to indicate any body of running water flowing through a clearly defined natural channel to progressively lower elevations. Every stream starts from a source. Some streams begin in the mountains, where snows or glaciers melt. Others begin as springs or **groundwater** that percolates onto the surface. On their way to the sea, streams collect water from rain and other streams (Figure 3).

Streams carry water, **sediment**, and **pollutants** from higher elevations downstream to lakes, estuaries, and oceans. This movement provides functions essential for maintaining life, such as cycling nutrients, filtering and diluting contaminants from **runoff**, absorbing and gradually releasing floodwaters, maintaining fish and wildlife habitats, recharging groundwater, and tempering **stream flows** (Federal Interagency Stream Restoration Working Group, 1998).

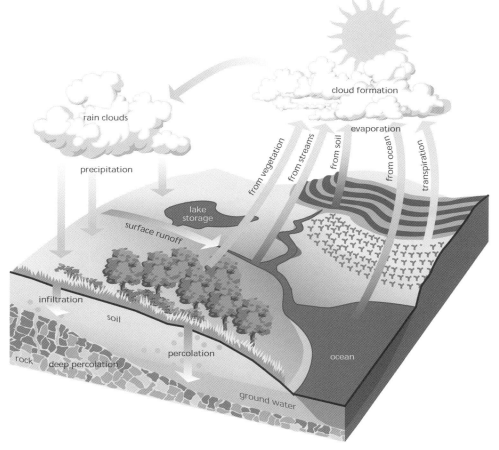

Figure 3. The hydrologic cycle consists of the movement of water from precipitation to surface water and groundwater, to storage and runoff, and eventually back into the atmosphere through **evaporation** and **transpiration.** (Federal Interagency Stream Restoration Working Group, 1998)

Watersheds

Every stream has a watershed, which is the total area of land from which water drains into a stream or other body of water. The watershed can also be called the **drainage basin** or catchment area. The boundary of a watershed is defined by ridges, hilltops, or less prominent surface features away from which water will **flow** and converge toward a common low point. All water flowing in the watershed will work its way toward the lowest point in the watershed. When all of the high points along the edge of a watershed are connected by a line, that line demarcates the boundary of the watershed. Someone standing by a stream at the bottom of a valley might look up and see hills that mark the edge of the drainage area. All of the land that slopes downward from the hills towards the stream is part of the watershed (Figure 4).

To visualize this concept, think about the rim of a bathtub. When water falls anywhere inside the tub, it runs downwards to the lowest point, which would be the drain. Water that falls outside the tub onto the floor makes its way to another low point in the bathroom. The rim of the

For assistance in delineating a watershed, please visit *www.nh.nrcs.usda.gov/technical/topomaps.html* or see Appendix A. To view, purchase, and learn to use maps and aerial photographs on-line, please visit *http://mapping.usgs.gov/*.

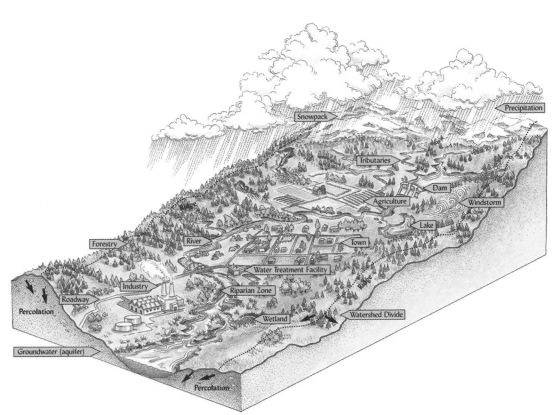

Figure 4. The boundary of a watershed corresponds to the tops of surrounding ridges, hills, or mountains. (USDA Forest Service figure)

bathtub forms the boundary between two watersheds.

To understand all the forces affecting a river or stream, it is important to consider everything that is taking place in the entire watershed. Therefore, the first step in a stream enhancement project is to identify the watershed. This task can be accomplished by contacting a local planning office for assistance or by using a topographic map (Figure 5; also see Appendix A). The thin brown **contour lines** on a US Geological Survey topographic map describe the rise and fall of the land. Each line represents a specific

elevation. Topographic maps also portray both natural and manmade features. They show and name natural features including mountains, valleys, plains, lakes, rivers, and areas of vegetation. They also identify human constructions, such as roads, boundaries, transmission lines, and major buildings. Topographic maps can also indicate drainage patterns and the direction of water flow on the landscape. Water flows down slope and perpendicular to the contour lines. V-shaped contours indicate a drainage line with the V pointing upstream. A continuous blue line indicates a **perennial** (constantly flowing) stream, whereas

Figure 5. Each contour line on a topographic map, such as on this section of a USGS map, represents a specific elevation above sea level. (US Geological Survey figure)

a dashed line followed by three dots usually indicates an **intermittent** (seasonal) stream. Aerial photographs can also be used to examine a watershed area.

Watersheds occur at multiple scales. They range from large river basins, such as watersheds of the Mississippi River, Columbia River, and Chesapeake Bay, to watersheds of small streams that might measure only a few acres in size. The size of a watershed depends on the topography of the landscape around the stream or river as it flows from the **headwaters** to the mouth. Even puddles have watersheds that exist inside the watershed of a larger stream or river.

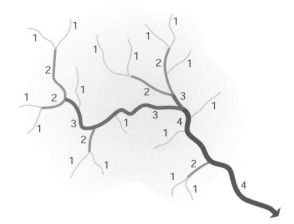

Figure 6. Stream ordering is used to classify individual channels in a drainage network. (Federal Interagency Stream Restoration Working Group, 1998)

The Need for a Watershed Perspective

Everyone lives and works in a watershed. Humans influence what happens in watersheds — good or bad — by how they treat the natural resources. Management decisions for a single resource in the watershed often affect other resources and the ecological functions of that watershed. What happens in small watersheds also affects the larger watersheds downstream. Efforts to enhance streams should consider all the activities in the watershed, including land-use practices that influence the waterways.

Examining land uses around the stream helps to determine the watershed characteristics. Is it grassland, forested, urban, or agricultural? Is the land around the waterway mostly paved, covered with vegetation, or exposed soil? Most watersheds have a mix of several different land uses. The key is determining how these land uses affect the stream. Assistance in determining the influence of surrounding land uses on streams can be

> **Find contact information for map resources and government assistance in the Watershed Stewardship Resources document available on-line at *www.iwla.org/sos/resources.***

Tip Box

Stream ordering is a technical method of classifying the hierarchy of natural channels in a watershed (Figure 6). A diagram of a river system looks very much like the branches of a tree. The smallest twigs of the river tree are the small streams where the river begins. These uppermost channels in a drainage network (headwater channels with no upstream tributaries) are called *first-order streams*. A *second-order stream* is formed below the confluence of two first-order channels. *Third-order streams* are created when two second-order channels join. Two third-order streams create a *fourth-order stream*. So what would result from the confluence of a first-order and second-order stream? The correct answer is a second-order stream, until it merges with another second-order stream, creating a third-order stream.

Figure 7. Land-use patterns in urban areas are usually marked by a high percentage of paved surfaces which impede the infiltration of surface water. (Jerry N. McDonald photograph)

obtained through government agencies and by using aerial photos, topographic maps, land-use maps, and watershed planning reports and assessments.

Learning about the watershed of a stream is critical to understanding potential sources of degradation of that stream. For example, if a watershed has many paved surfaces, much of the watershed is **impervious** to water (Figure 7). Rainwater cannot soak into the ground; instead, it will run off quickly, sometimes collecting pollutants. In another watershed, if livestock are allowed to wade into a stream, they may cause degradation by depositing manure, trampling stream banks, and eating streamside vegetation. Streams that are bordered by cropland and lack protective strips of vegetation, or **buffers,** could suffer from the overuse of pesticides and fertilizers that are carried into the stream by runoff (Figure 8). Each waterway must be examined on a case-by-case basis, incorporating all the land-use activities within its watershed.

Stream enhancement projects that do not consider current and future land-use patterns have few chances of

succeeding. Stabilized banks must accommodate the water behavior created by the upstream conditions in the watershed. Urban streams flow differently than rural streams because the greater amount of impervious surface affects the timing, speed, and quantity of incoming water. Successful enhancement projects should execute watershed-specific solutions based on careful planning and study of the entire watershed system.

Stream Types

Streams, or channels, form in a variety of ways. Water moves down slope and concentrates in low areas, forming small stream channels. Channels that only carry water when and immediately after it rains are referred to as **ephemeral streams** (Figure 9). Downstream from ephemeral channels are intermittent streams, which carry water only during wet times of the year. These streams may be partially supplied by groundwater, and they disappear

Figure 8. Highly mechanized agriculture often requires large inputs of pesticides and fertilizer. (Bureau of Reclamation photograph)

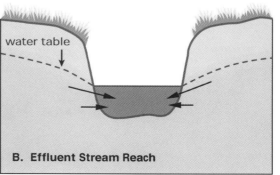

Figure 9. Ephemeral streams flow only during or immediately after periods of precipitation. (Federal Interagency Stream Restoration Working Group, 1998)

when groundwater levels drop. Farther downstream, where groundwater levels are large enough to sustain stream flow throughout the year, perennial streams develop.

There are two primary types of streams in relation to groundwater: **effluent**, or *gaining* streams; and **influent**, or *losing* streams (Figure 10). Gaining streams receive water from groundwater sources; losing streams supply water to groundwater **aquifers**. In times of drought, perennial streams continue to flow because they are receiving water from the groundwater aquifers. In more arid regions of the United States, losing streams are dry for part of the year because water is absorbed back into groundwater aquifers.

Stream Stability

Streams naturally strive to achieve a balanced state of **dynamic equilibrium**, where the amount of water and sediment leaving the stream is equal to the amount entering the stream. Although stable streams do achieve dynamic equilibrium over time, it is natural for the amount of water and sediment coming into a stream section or **reach** to periodically fluctuate as streams naturally flood and deposit sediment along the banks. Naturally stable streams are able to transport the sediment load deposited into the stream over the long run.

Streams constantly change their course to accommodate the natural and human-induced changes in the watershed that add or remove water and sediment from the system. Changes in the watershed, such as clearing forested land, add sediment to the stream, causing changes in stream channel shape, depth, and other characteristics. Soil types, geology, vegetation, topography, and climate also affect channel **morphology** (shape).

It is important to understand the behavior of naturally stable streams in order to identify signs of stream instability. Streams that are in dynamic equilibrium

Figure 10. Influent streams (A) lose water to underlying aquifers. Effluent streams (B) are continuously recharged by groundwater reserves. (Federal Interagency Stream Restoration Working Group, 1998)

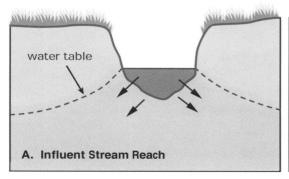

A. Influent Stream Reach

B. Effluent Stream Reach

have a balance between sediment **erosion** and sediment **deposition** (the accumulation of soil particles on the streambed). When the natural balance of a stream is disrupted, excessive erosion or sediment deposition occurs. Excessive bank erosion can sweep large quantities of soil off banks, causing deposits of sediment downstream. **Bank failure** occurs when the bank materials are not strong or stable enough to resist gravity. This usually happens after a rain event that causes the soil — heavy with water — to slump off at weak points in the bank. Gravel or sand bars that frequently change locations often indicate an unstable system trying to adjust to a high influx of coarse sediment.

The sediment deposited into the channel from **stream corridors** can suffocate fish, smother spawning beds, kill aquatic insects, and degrade water quality. Sediment also blocks sunlight needed by aquatic plants and can contain pollutants, such as pesticides, fertilizers, and heavy metals, which cling to soil particles. If erosion is not controlled, valuable farmland can be lost, water quality can be degraded, and human health can be endangered.

Stream instability occurs when erosion causes the channel to be lowered or to deepen (**stream channel degradation**) or excessive sediment deposition causes the channel to rise (**aggradation**). As the streambed lowers it can undercut the stream bank, causing instability (Figure 11).

Figure 11. Exposed tree roots indicate that this stream bank is being eroded through the process of undercutting. (Izaak Walton League of America photograph)

Stream stability is affected by four basic factors:

- Sediment load (the amount of sediment)
- Sediment size
- Stream **slope**
- Stream flow or **discharge**

The amount and kind of sediment carried by a stream largely determines its characteristics, including size, shape, and profile of the stream. The relationship between these factors is a balance between the product of the sediment load (the amount of sediment) and sediment size and the product of stream slope and stream discharge. The balance can shift depending on changes in sediment size, stream slope, and the quantity of the two. If one variable changes, one or more other variables must increase or decrease proportionally in order to maintain a state of equilibrium. A change that causes imbalance will make the stream degrade or aggrade until balance is restored. Changes in these variables affect channel width, channel depth, slope, discharge or flow, **velocity**, sediment size, and **channel roughness** (Federal Interagency Stream Restoration Working Group, 1998).

Stream Flow Behavior

The size and water flow of a stream are directly related to its watershed area, soils, geology, **hydrology**, plants, and climatic conditions. Stream flow, or discharge, is the volume of water moving through the channel over time, and it is usually expressed in cubic feet per second (cfs). The terms flow and discharge are used interchangeably throughout this publication. The intensity, distribution, and seasonal effects of precipitation influence stream flow. Stream flow consists of **storm flow** and **base flow**. Storm flow is the water from precipitation that rolls over the land or through the shallow subsurface ground materials to the channel after a storm and may cause flooding. Base flow, the contribution from the groundwater, moves slowly through the ground before reaching the channel and sustains stream flow during periods of little or no rainfall (Federal Interagency Stream Restoration Working Group, 1998).

Bankfull Discharge

The most important stream flow characteristic that determines the shape of the channel is the **bankfull discharge,** the volume of water flowing in a stream that is at **bankfull stage,** and stream enhancement projects will need to acknowledge, understand, and accommodate this volume of water. In its simplest and most easily understood form, that of a stable stream in an alluvial valley, bankfull stage is the elevation of the upper surface when that stream has filled its banks and is ready to begin spilling out onto its **floodplain.** Bankfull stage is also called **bankfull,** bankfull elevation, and bankfull level. Bankfull stage is sometimes equated with a legal designation called the **ordinary high-water mark**, below which streams have

greater protection under federal and some state laws. Bankfull discharge is the volume of water, usually expressed in cubic feet per second (cfs), that a stream is carrying when it reaches the bankfull stage. Bankfull discharge is also called bankfull flow, channel-forming discharge, dominant discharge, and effective discharge. In unstable streams or in settings other than alluvial valleys — deeply **entrenched streams** or streams in urban areas or other highly developed watersheds, for example — the recognition and determination of bankfull stage and bankfull discharge become more complex and sometimes require considerable experience or professional assistance to identify and properly calculate.

On average, bankfull discharge occurs approximately once every 1.5 years. In other words, each year there is about a 67 percent chance of having a bankfull stream event. In addition to being a discharge that occurs occasionally, bankfull discharge is also a theoretical flow of water that, if constantly maintained over a long period of time, would produce the same channel shape as would all the high and low flows that naturally happen over time. While a large storm event, such as a 100-year flood, may cause vast amounts of erosion and dramatically change the location and shape of a channel, the channel is more affected over time by smaller flood events that occur more often. Bankfull discharge has enough power over 50 to 100 years to shape and maintain a channel.

When a stream is in equilibrium with its channel, its banks should be stable and contain the water when flow is at or below bankfull discharge, for it is this flow that carries out most of a stream's erosion, that transports most of a stream's sediment

over time, and that is most responsible for forming the size, shape, and other characteristics of a stream's channel. Because the shape and other features of a stream's channel are determined by bankfull discharge, it is important to have this flow in mind when designing any stream enhancement or restoration project.

Stream flow that rises above the bankfull stage can cause the stream to flood into an area called the floodplain. This flooding would happen in a stable stream in an alluvial valley, but if changes had occurred in the watershed that had rendered the stream system unstable, other results might occur instead. An increase in the erosion rate of the stream, for example, will cause the stream to erode a deeper channel and, as a result, waters of the natural bankfull stage might not be able to reach the top of the stream's banks and spill over onto the floodplain. A stream that cannot reach the floodplain when flowing at bankfull discharge is either an **incised** or entrenched stream (Figure 12). These streams are more likely to erode their banks during bankfull and flood events, and the elevation of the bankfull stage might be marked with a **scour** line on the stream's banks or other soil and vegetation features. If the stream is not entrenched, it is safe to assume that bankfull stage is located at or near the top of the bank (Figure 13). Methods to determine bankfull flow are discussed in Chapter Three.

Floods

Floods are a natural cycle of river ecology. Animal and plant communities have spent millions of years adapting to floods. **Riparian zones** depend almost exclusively on flooding cycles for their existence. Many fish wait until the first sign of the annual spring flood to start breeding. Insect larvae wait for flooding to begin to lay eggs, hatch, or metamorphose. Flooding provides new food sources for stream inhabitants by flushing terrestrial invertebrates into the water. In addition, flooding results in increased fertility for the stream as nutrients (such as nitrogen and phosphorus) are washed out of the soil and converted to food sources for aquatic life.

On the other hand, floods can cause damage to streams. For example, while nutrients support the base of the **food chain**, too many nutrients can damage an ecosystem by creating excess **algae** that depletes oxygen levels in the stream. Streams without adequate stream bank vegetation lack the structure to hold soil in place during floods and can get overloaded with nutrients and sediment.

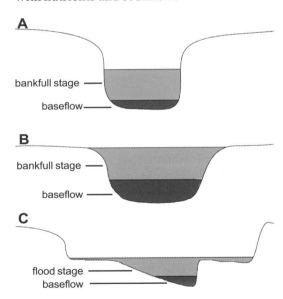

Figure 12. (A) In an entrenched stream, bankfull discharge is not able to spill over the banks onto a stream's floodplain. Therefore, the bankfull stage is below the top of the banks. In a stable stream in an alluvial valley, (B) bankfull stage is often located at the top of the bank but (C) flows greater than bankfull discharge are able to spill over the banks onto the floodplain. (Izaak Walton League of America figure)

Figure 13. A river in equilibrium at (A) bankfull stage and (B) soon after reaching flood stage. (John A. Conners photographs)

It is important to plan for floods. Flood experts rank the size of floods in statistical terms that represent how frequently a certain amount of water will flow through a stream channel over time. Flood frequency is a period of time that occurs between floods of the same or greater levels. For example, a flood that happens approximately once every 10 years is called a 10-year flood. In any given year, a 10-year flood has a 1-in-10 chance of occurring. Terms such as a 10-year flood, 25-year flood, and 100-year flood are based on historical data. A 100-year flood statistically has a 1-percent chance of occurring during any given year. A 20-year flood has a 5-percent chance of occurring in any year. The frequency with which these floods occur fluctuates due to many factors, such as years of high rainfall and changes in land use. For example, the volume of water that used to be the 100-year flood might begin to occur every 5 or 10 years in an urbanized watershed due to an increase in the amount of impervious surfaces. Stream enhancement projects should ideally consider the 100-year floodplain area. Enhancing the **vegetative** structure in this area will reduce the amount of pollution that washes into the stream during flood events.

A stream enhancement project needs to consider both the bankfull discharge and normal flood frequency. A stream enhancement project designed to accommodate more water or materials than are available may cause the steam to fill in and deposit sediment in bars. As more sediment is deposited into a stream, the streambed will become shallower, forcing water to spread out and thereby widen the stream. Widening causes more bank erosion and still more sediment deposition. Sediment deposition smothers life forms and fills-in **pools** needed by larger fish. A project that underestimates bankfull discharge and flood frequency will cause an increase in local flooding and bank erosion during a storm event.

In-Stream Features

The processes that move water, materials, and organisms through a watershed determine the look of a stream. Stream features provide varying sources of

food and habitats and encourage plant and animal diversity. The presence or absence of various stream features offers clues about the stability of a stream. The stream features can be examined from three different views:

- The stream **dimension** (the lateral or cross-section view)
- The stream **pattern** (a top-down or plan view)
- The stream profile (the longitudinal view)

The following subsections identify and introduce terms associated with in-stream features that are useful in assessing stream health and planning a stream enhancement project (Figure 14).

Stream Dimension

The dimension of a stream is its cross-sectional area (width multiplied by average depth). The sloped bank is called a **scarp** and the deepest part of the channel is called the **thalweg** (Figure 15). The width of a stream generally increases in the downstream direction. Stream width is a function of the amount of water that passes through the channel without spilling over the banks, the amount of sediment available to be picked up by the flow, and stream bank vegetation and bed materials. A **channel cross-section** will widen with increased discharge and narrow with an increased sediment load (Interagency Federal Stream Restoration Working Group, 1998). Increasing a stream's discharge (by adding more impervious surface in the watershed, for example) can also increase the channel depth and slope. A stream's cross-sectional area can be used to estimate the water discharge — information that is needed to design a stream enhancement project. Stream banks

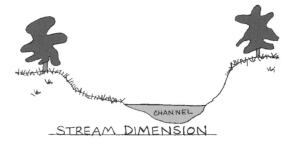
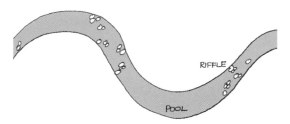
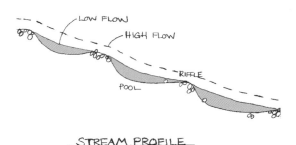

Figure 14. Dimension, pattern, and profile are three features of a stream that are used to monitor stream health and plan appropriate restoration projects. (Izaak Walton League of America figure)

that are either too deep or too wide for the bankfull discharge will not persist over time. If left alone, the stream will create a more appropriate channel through periods of erosion and deposition.

Another aspect of stream dimension is channel roughness. Channel roughness describes the amount of material in the stream providing friction against the movement of water. Friction between the stream and the bed and banks is important because it slows water flow and helps to create turbulence. This turbulence adds dissolved oxygen to the water that is vital for fish and aquatic insects.

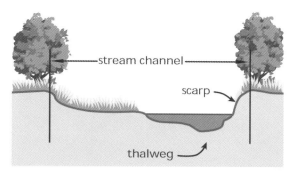

Figure 15. The dimension of a stream channel can be used to determine the water discharge. (Federal Interagency Stream Restoration Working Group, 1998)

Stream Pattern

Redesigning a stream pattern is beyond the scope of a volunteer-driven stream enhancement project, but understanding patterns of water flow is important when considering stream enhancement projects.

Stream pattern refers to the *plan view* of a channel as seen from above. Water does not typically follow a straight course unless it encounters extremely erosion-resistant rock or human-made obstructions or channels (Figure 16). Natural streams are rarely straight over a distance of more than 10 times the width of the channel. The curvature of a stream is called **sinuosity** and can be described as straight, meandering, or braided. The more strongly curved the shape of a channel, the more sinuous it is. The channels of braided streams are less sinuous than those of meandering streams and possess three or more subchannels, each separated from the others by islands or sand or gravel bars, within the channel proper (Figure 17).

Streams adjust their pattern so that the water has to do as little work as possible. Streams naturally create **meanders** to slow water flow. Slow-moving

water has less power to erode than fast-moving water. Streams flowing through soft soils can create meanders more easily and flow more slowly than streams flowing through erosion-resistant rock. As water moves around a meander, the greatest force of the water hits the downstream portion of the bend, causing significant bank erosion on the outside of meanders. A corresponding amount of deposition typically occurs on the inside of meander bends, resulting in the lateral migration of the entire stream channel.

Water flow can create several types of topographic features as the channel migrates in response to changes in the watershed upstream (Figure 18). Recognizing and understanding these features helps in determining what is happening in the stream and which enhancement techniques could benefit the stream.

A meander is a circuitous winding or bend in the river. A **chute** is a new channel formed across the narrow part of a meander that bypasses the meander and allows faster stream flow. A **meander scroll** is a trail of distinguishable sediment marking

Methods to determine cross-sectional area and discharge for stream enhancement projects are discussed in Chapter Three.

Figure 16. Stream channelization and armoring eliminates natural sinuosity and results in ecological decline. (Izaak Walton League of America photograph)

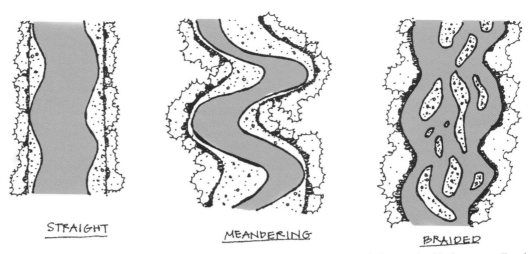

STRAIGHT   MEANDERING   BRAIDED

Figure 17. Stream channels are characterized as having a straight, meandering, or braided pattern. (Izaak Walton League of America figure)

former channel locations. An **oxbow** is a meander that is severed from the channel once a chute is formed, and an **oxbow lake** is the body of water created after sediment blocks the oxbow inlet. Natural **levees** are low ridges of sediment that build up along the bank of some streams that flood. They form as sediment-laden water spills over the bank, encountering a sudden loss of depth and velocity, causing coarser-sized sediment to drop out of suspension and collect along the edge of the stream. **Splays** are delta-shaped deposits of coarser sediments that occur when a natural levee is breached. Natural levees and splays can prevent floodwater from returning to the channel when floodwaters

recede. **Point bars** form on the inside of meanders, where water flows more slowly and lightweight sediment falls out of suspension. **Island bars** form in wide streams, where water flows too slowly to transport the available sediment or where the channel material lacks the cohesiveness to maintain distinct banks along the stream margins.

Stream Profile

The **longitudinal profile** of a stream refers to its longitudinal slope. **Channel slope** (vertical rise divided by horizontal run) generally decreases as the stream flows down the watershed. The size of the sediment on the streambed also decreases in the downstream direction. For example,

Figure 18. This plan view illustrates the diverse landforms and deposits that can form in the floodplain of a stream. (Federal Interagency Stream Restoration Working Group, 1998)

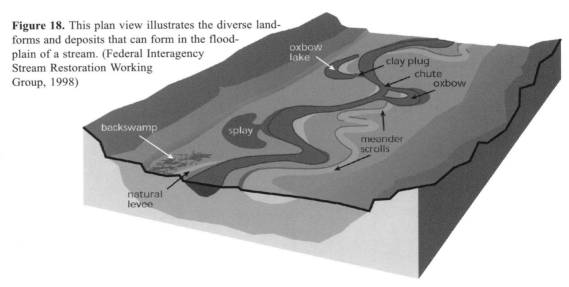

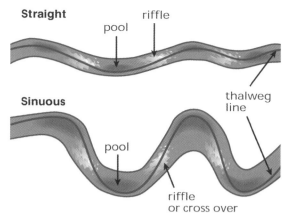

Figure 19. Pools and riffles are natural features of both straight and sinuous stream channels. (Interagency Stream Restoration Working Group, 1998)

streams in mountainous areas often have beds of rock or boulder while coastal streams tend to be sandy. Channel slope is inversely related to sinuosity. This means that streams that run over steep terrain are relatively straight (low sinuosity) compared to streams in flat lands, which are usually very curvy (high sinuosity). As stated earlier, sinuous channels dissipate the water's energy and decrease erosion.

Natural streams have sequences of **riffles** and pools — or **steps** and pools — in the stream profile, which maintain channel slope and stability. Riffles are found before and after meander bends, where water bubbles over rocks (Figure 19). The riffle is a streambed feature with gravel or larger size rocks where the water depth is relatively shallow and the slope is steeper than the average slope of the channel. As water flows over the rocks, oxygen is infused, creating an oxygen-rich environment for fish and aquatic insect larvae. Rocks also provide shelter for aquatic insects and spawning beds for fish, and trap organic matter that becomes food for the aquatic ecosystem. In sandy or muddy-bottom streams found in coastal areas, snags and fallen trees serve a purpose similar to that of rocks in a riffle.

A pool has a flat slope and is much deeper than the average depth of the stream. Pools are located on the outside of meanders and provide deep-water habitat for larger fish. Burrowing insects, such as some species of mayflies and dragonflies, can be found in pools. At times of low flows, pools collect depositional materials and riffles provide scour materials. At high flows, however, the pool scours and bed materials deposit in the riffle.

Step/pool sequences are found in steep streams. Steps are vertical drops often formed by large boulders, bedrock, downed trees, etc. Deep pools are found at the bottom of each step. The step controls the water flow down the steep slope and the pool dissipates energy. Pool and riffle structures can be reestablished in streams. This task, however, requires the assistance of natural resource professionals who have experience with in-stream restoration.

It is possible to determine the slope of a stream by noting the presence or absence of steps and pools. The spacing of steps and pools gets closer as the channel slope increases. In high mountain areas with lots of pools and waterfalls, the **gradient** usually is greater. If the stream meanders a great deal, the stream's gradient probably is low. Identifying and monitoring all stream features will provide information needed to develop a stream enhancement project.

Stream Corridor Features

Stream corridors include streams, their channels, their floodplains, and the adjoining upland fringe. Stream corridors are critical ecosystems in our living environment. They are complex systems made up of air, water, land, plants, and animals, all of which interact and depend

upon each other. These systems perform many ecological functions such as moderating the amount of water in streams, storing water, removing harmful chemicals from the water, and providing habitat for aquatic and terrestrial plants and animals. For example, floodplains and **terraces** hold water overflow, buffer zones provide wildlife food and habitat, and vegetation along the banks holds soil in place and filters water as it flows into the stream.

Floodplains and Terraces

The floodplain is the streamside land that periodically gets inundated by the river's floodwaters. Floodplains perform several important functions: they store floodwaters temporarily, filter water quality, provide habitat for aquatic wildlife, and create opportunities for recreation, such as picnicking or fishing. During periods of high water, floodplains serve as natural sponges, slowly releasing floodwaters. Floodplains also play an important role in recharging groundwater supplies.

Stream corridors often contain one or more topographic terraces, or benches, above the floodplain. Terraces are remnants of old floodplains that have been abandoned (sometimes thousands of years ago) as a stream's dynamic equilibrium evolves (Figure 20). Changes in the amount of stream flow or sediment can cause channels to deepen and widen. When this happens, the channel will continue to widen over time and establish a new floodplain at the lower elevation, making the abandoned floodplain an upland terrace (Figure 21).

Buffer Zones

Forests, shrub lands, or grasslands along a stream are referred to as buffer

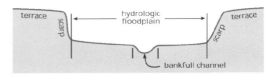
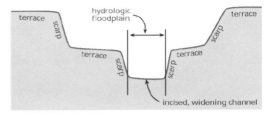
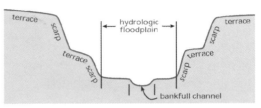

Figure 20. Changing watershed conditions and shifts in the stream's dynamic equilibrium result in the evolution of the stream channel. (Federal Interagency Stream Restoration Working Group, 1998)

zones or strips because they buffer the effects of land uses on streams. Forested buffer strips trap and filter pollution in the forest floor's organic layer. The velocity of runoff through a vegetated area is slower than over bare soil or pavement, and therefore less erosion occurs. Vegetation also holds water and then slowly releases it. Microorganisms that live on tree roots help breakdown and take up pollutants. Trees with deep root systems can absorb toxins from the groundwater (Figure 22).

Older forests filter better than younger forests because the organic layer is thicker and the root system is more dispersed.

Various natural and human features, including **wetlands**, riparian forest buffers, oxbow lakes, residential or commercial developments, and farming practices, are common along stream corridors. Wetlands help slow and absorb floodwater

Figure 21. This incised stream, now dammed and flooded, has at least three terrace levels (abandoned floodplains) within its watershed. (Bureau of Reclamation photograph)

food, and migration corridors. Migratory waterfowl and songbirds use forested stream corridors and wetlands during their annual migration, and floodplain trees serve as important roosting and nesting habitat for raptors such as the bald eagle.

Stream Banks

Stream banks are the sloping ground that border a stream and confine the water in the natural channel when the water level is normal. Banks are called right or left as viewed facing the direction of flow. Stream banks have vegetation and soil characteristics distinctly different from surrounding uplands, which can support higher levels of species diversity, species densities, and rates of biological productivity. Because

and prevent rapid runoff. Wetlands also provide spawning and rearing habitat for a variety of fish species, which provide food for waterfowl, other wildlife, and people. Wetland plants help trap and filter sediment, and microbes on wetland plant roots actually break down pollutants. Wetlands can be wet all or some of the year and are characterized by **hydrophytes** (water-loving vegetation) and **hydric soils** that are often gray or mottled in appearance. Many cities have built artificial wetlands to reduce water treatment costs. Studies of heavily polluted waters flowing through Tinicum Marsh in Pennsylvania, for example, have shown significant reductions in phosphorous and nitrogen. The value of the ability of Georgia's 2,300-acre Alcovy River Swamp to take pollutants out of the water is worth more than one million dollars a year.

Buffer zones also provide the food and shelter wildlife need to survive and reproduce. Nearly 20 percent of all vertebrate species rely on the buffer zone during their life cycle. Healthy riparian zones create a vegetated transitional area between streams and upland habitats, providing shelter,

Figure 22. Buffer zones filter runoff from surrounding land uses. (Pam Cullen photograph)

stable, vegetated stream banks provide many benefits to wildlife and water quality, many enhancement projects concentrate on bank stabilization.

Vegetation

Streamside vegetation plays a critical role in protecting stream banks and water quality, and providing vital habitat for aquatic and semi-aquatic animals. Trees and woody vegetation draw water from the stream banks, helping to improve **bank stability**. In addition, tree roots that hang from stream banks provide places for fish and other animals to hide from predators. Long roots growing out into the water also create eddies and ripples that help oxygenate the stream. Leaves falling from trees provide food for aquatic insect larvae that eat decaying leaves. Finally, trees provide shade for the stream, keeping water temperatures cooler, which allows the water to hold more dissolved oxygen for fish and aquatic insects.

Observing and understanding how fallen trees and limbs function in a stream is important to determining the appropriate enhancement measures.

Although fallen trees can cause erosion and bank failure if they deflect water onto a bank with enough force, they play a critical habitat role for the aquatic species. They provide cover for fish, serve as a medium where algae and plants can grow, and increase the roughness of the channel, which slows water flow. Sometimes trees that fall across the channel can block flow and catch other pieces of floating wood. In some cases, this can present a recreation hazard or cause localized erosion problems

See Chapter Three for more information on evaluating riparian vegetation and Chapter Four for guidance on how to select appropriate vegetation for an enhancement project.

when floodwaters scour around the logjam. Therefore, it is important to work with local experts to fully understand how a fallen tree or log is functioning in the stream before deciding if it is necessary to remove it.

Variations in Plant Communities

The structure of plant communities varies throughout the stream corridor. The distribution of these communities is based on the soil and hydrology of the specific area. Some riparian plant communities, such as willows and cottonwoods, depend on flooding for nutrients and are common in broad, flat floodplains. Upland plants, such as those in a forest on a moderate to steep slope in the eastern or northwestern United States, might come close to bordering the stream and create a canopy that covers the channel.

Differences in climate across the country affect watershed vegetation and natural stream form. More vegetation grows in the climates of the East and Northeast than in other regions. Vegetative groundcover reduces erosion by intercepting raindrops and holding rainwater in roots and soils, slowly releasing it during dry periods. The presence of vegetation can influence whether the stream flows seasonally (intermittent) or year-round (perennial).

Arid to semi-arid areas of the West contain sparse vegetation, such as chaparral or coastal scrub, which produce less-developed soils. Streams with these easily eroded, fine-textured soils are more likely to be incised than streams in areas with more vegetation and more well-developed soils. Similarly, stream banks in humid, subtropical climates are more likely to be narrower than those in arid regions because of the presence of more vegetation.

Plant communities play a significant role in determining stream corridor condition and potential for enhancement. Thus the type, extent and distribution, soil moisture preference, elevation, species composition, age, vigor, and rooting depth of vegetation are all important characteristics to discuss and investigate with natural resource professionals when planning a stream enhancement project (Federal Interagency Stream Restoration Working Group, 1998).

Natural and Human-Induced Disturbances

Human activities and natural events can bring changes to stream corridor structure and function. Disturbance can be a natural process. The health and equilibrium of streams are dependent upon natural disturbances, such as floods, which help to maintain important ecological processes. Human activities can disrupt these natural disturbances that allow watersheds to maintain their dynamic equilibrium.

Natural events that disturb the structure and function of stream corridors include hurricanes, tornadoes, fire, lightening, volcanic eruptions, earthquakes, insects and disease, landslides, temperature extremes, and droughts. The ecosystem's response to these disturbances varies according to its relative stability, resistance, and resilience. In many instances, ecosystems recover with little or no need for supplemental enhancement work. Enhancement efforts, however, can help stabilize disturbed land and help put streams back into a balance with the surrounding watershed.

Human-induced disturbances brought about by land-use activities have the greatest potential for introducing enduring changes to the ecological structure and functions of stream corridors. A single disturbance might trigger a variety of

sequential disturbances that permanently alter one or more characteristics of healthy systems. It is important to base stream enhancement projects on an understanding of how various disturbances affect stream corridors and how the system responds to those disturbances.

Land-use activities can disrupt physical, biological, or chemical processes in a watershed and stream corridor. Chemical pollution, for example, can be introduced through many land uses, including agriculture (pesticides and nutrients), urban development (municipal/industrial/household waste contaminants), and mining (acid mine drainage/introduction of heavy metals). Most of these chemicals are **toxic** to aquatic life and degrade water quality (Federal Interagency Stream Restoration Working Group, 1998).

Chemical disturbances can originate from **point** and **non-point** sources of pollution. **Point source pollution** stems from one location, such as the end of a pipe, and is relatively easy to control. **Non-point source pollution** comes from various places across the surface and subsurface of the land. Secondary effects, such as agricultural chemicals attached to sediments and increased soil salinity, frequently result from physical activities (irrigation or heavy application of herbicides). These activities are best controlled at the source rather than by treating the symptoms within the stream. The Environmental Protection Agency has developed strict regulations that control point source pollution, which now makes non-point source pollution the greater concern (Federal Interagency Stream Restoration Working Group, 1998).

Biological disturbances, such as improper grazing management or recreational activities, can result in the

establishment of **exotic species** — species of microbes, plants, and animals that are alien or non-native to an ecosystem. Some exotic species can create widespread, intense, and continuous stress on native biological communities. They can introduce diseases and compete with native species for moisture, nutrients, sunlight, space, and food. Exotic species that have no natural predators in their new environment can proliferate, take over an area, and crowd out native species. They can even detract from the recreational value of streams by creating a dense, impenetrable thicket along the bank (Federal Interagency Stream Restoration Working Group, 1998). Exotic species that create economic or environmental harm to ecosystems or to human health are referred to as **invasive species**.

Physical disturbances of the stream and its watershed are caused by activities such as flood control, forest management, road building and maintenance, agricultural tillage, and irrigation, as well as urban development. Loss of **stream buffers**, floodplains, wetlands, and stream bank habitats can affect the infiltration and movement of water and alter the timing and magnitude of runoff events. Common disturbances to stream corridors include removing buffer vegetation, soil exposure, soil compaction, and construction of impervious surfaces. Eliminating shade trees in buffer areas increases water temperature during summer and decreases it in winter. Reduced vegetative cover can increase soil compaction and decrease the depth and productivity of topsoil.

Either individually or in combination, human disturbances place stress on the stream corridor that alters its structure and impairs its ability to perform key ecological functions. Actions needed to enhance stream corridors are best determined by understanding the natural evolution of streams, which disturbances are stressing the system, and how the system responds to those stresses. Each stream has unique characteristics and requires a unique enhancement strategy. Local natural resource professionals can help provide the insight into stream conditions in your community. Community members can play a key role in planning stream enhancement and stewardship projects.

Chapter Two

Assessing the Watershed

A **watershed assessment** is the vital first step in planning a stream enhancement project. A thorough assessment provides the information needed to choose appropriate sites and enhancement techniques. When planning a stream enhancement project, it is important to grasp all the factors that may affect the watershed, currently and in the future. Some of these factors include the stream's water quality, its flow capacity, the suitability of its habitat for fish and wildlife, the primary uses of the stream by people and animals, and past alterations to the stream. For example, if a goal of the stream enhancement project is to bring back a native fish species to the stream, it is important to assess whether the planned improvements will adequately achieve the goal. It might be that the stream's water quality will not support the fish or that stormwater flows are too high. When the scope of the stream enhancement project is greater than a community's capacity to achieve the goals,

it is recommended that the goals of the initiative be revised (Figure 23).

Establish a Technical Team

Prior to beginning any work in a watershed, the Izaak Walton League advises that a group establish a technical team of experts within the community to assist with the overall stream enhancement project. The technical team needs to be on board before the project can get off the ground, so begin contacting potential members as soon as you feel comfortable. The experts you find — whether in government agencies, schools and universities, consulting firms, or other organizations — will depend on who is locally available and who is willing to volunteer, work on a *pro bono* basis, and who will only participate if they are paid. Developing an interdisciplinary team of specialists will provide the knowledge, skill, ability, and professional judgment a project needs

Figure 23. In designing alternatives for stream bank erosion, it is important to assess the feasibility of addressing the cause of the problem (e.g., modification of land uses) or treating the symptoms (e.g., installation of stabilization structures). The degree of erosion illustrated here may be more than a community service project can handle. (Izaak Walton League of America photograph)

to succeed. Experts often have access to information needed for the watershed assessment, site selection, inventory, engineered design, permits, installation, monitoring, and maintenance. For more details on establishing a technical team, please see Appendix B.

Watershed Assessment and Site Inventory

There are two types of assessments that should be considered before starting a project: watershed assessment and **site inventory**. A watershed assessment involves examining the entire watershed by walking along the banks of the streams to determine their condition. A watershed assessment can identify sites that are priorities for enhancement. An assessment of the watershed also aids in the identification of **reference sites** or **refer-**

Start your assessment by contacting your local government to find out about their watershed or water resources programs.

ence reaches — stream segments with good bank stability, diverse riparian vegetation, and good water quality that can serve as models for the stream enhancement project (Figure 24). More information about reference reaches is available later in this chapter. Although many different watershed assessment techniques have been developed, most are designed to allow professionals or volunteers to quickly move through the watershed, cataloguing the conditions of each stream site therein.

Site inventories, on the other hand, are in-depth examinations of specific stream segments designed to gather a wide range of information about the water quality, hydrology and **hydraulics**, bank and bed stability, in-stream and stream corridor habitat, and other factors that determine overall stream health. This information is used to determine the most appropriate techniques for enhancing the site. Site inventories also provide baseline data about the stream that can be compared with future monitoring results to measure the success of a project. An in-depth site inventory also can help volunteer groups determine whether they can accomplish the scale of the enhancement project that will adequately address the needs of the stream. Site inventories are addressed in detail in Chapter Three.

The League recommends that groups prioritize stream sites based on severity of degradation and then work on projects in their order of priority, starting with the sites farthest upstream and working downstream. Understanding conditions upstream in the watershed is important because any structural changes made to

Figure 24. Walking through a watershed is an important first step in a thorough stream bank assessment. (Izaak Walton League of America photograph)

the stream could either increase erosion or deposition problems downstream or be destroyed by upstream forces. Ignoring problems upstream results in a waste of time and money and will discourage sponsors, supporters, and volunteers from participating in future enhancement projects.

Research Existing Information

Before conducting a watershed assessment, plan on investing the time needed to research any existing information on the stream or watershed. Local government agencies often conduct watershed assessments to help with planning efforts and to determine priority sites for their own enhancement projects. If the county or city planning or environmental offices have a watershed assessment and/or restoration plan, agency staff will be able to help your group determine appropriate site(s) for enhancement.

If these agencies do not have a watershed assessment or restoration plan, work with them to develop one. Many local government agencies are planning for future development projects and will be able to provide valuable information that

could affect the stream enhancement project, such as future changes in the quantity of stormwater that will flow into the watershed. Local universities, the chamber of commerce, environmental agencies, departments of highways or mining, and soil and water conservation districts might have information about the stream and activities that have altered it. Other good sources of historical information are long-time residents, libraries, or local nongovernmental organizations.

The watershed assessment typically begins by reviewing and interpreting maps and other written or visual resources. As mentioned in Chapter One, topographic maps show streams and elevation changes, and can be used to define the watershed boundary. In a natural setting, watershed boundaries — the line that separates one watershed from another — can be drawn by connecting the highest points in elevation that surround the system of streams. In urban areas, maps of the storm drain system from the city or county will reveal whether rain that falls in one watershed is transported through pipes into a different stream system. The US Geological Survey 7.5-minute topographical map can be used to determine drainage patterns and identify land areas that may be affecting the stream's health. Note where major features (such as cities, landfills, mines, highways, and forestry areas) fit into the watershed, since these are often sources of possible sedimentation and pollution.

In addition to the US Geologic Survey maps, the USDA Forest Service and the US National Park Service, Fish

and Wildlife Service, and Bureau of Land Management have resource-specific maps available for lands they manage and protect. Information on the watershed boundary, historical data, monitoring data, and other information on the stream also may be available on the Environmental Protection Agency's Surf Your Watershed web site at *www.epa.gov/surf.* Comparing older maps and photographs with more recent ones can provide valuable information on stream corridor changes over time.

It is important to evaluate each stream segment in the watershed. Unless otherwise specified by the watershed assessment method chosen, the length of each segment should be about 20 times the **active channel width** to collect information on average stream characteristics. The active channel width is the stream width at the bankfull stage. For example, if the channel width is 15 feet, each segment to be assessed should be 300 linear feet. However, if the conditions change dramatically along the stream within this length, break the area into smaller segments that share similar conditions. The bankfull stage of a channel can be difficult to determine, but it is worth becoming familiar with the concept and its field determination. Also, an estimate of the active channel width is adequate for many watershed assessment methods.

View, purchase, and learn to use maps and aerial photographs online at *http://mapping.usgs.gov.*

Figure 25. A reference reach may be similar to what the corridor was like prior to a disturbance. (Pam Cullen photograph)

Locate Reference Reaches

To best understand what the stream corridor should look like after enhancement, it is important to find a reference stream or part of a stream, also known as a reference reach or site. A reference reach represents what the degraded site could look like had it remained stable. A reference reach may not be pristine, but it should represent what is reasonably attainable over time (Figure 25).

It is very important that the reference reach — whether it is located in the same watershed with the site to be improved or not — shares certain characteristics with the project site. In order to qualify, a reference reach must be similar to the project site in size, stream type, location in the landscape, surrounding land use, etc. For example, it would not be practical to use a bedrock stream in the mountains as a model for a project site located on the coastal plain. One way to make sure that the reference reach matches the stream site to be enhanced is through the use of stream classification (discussed in Chapter Three). Talk with the local department of natural resources and the local Natural Resources Conservation Service office to request their assistance with identifying reference reaches. The watershed assessment will help to identify high quality areas within the watershed that can serve as a reference.

Examples of Watershed Assessments

There are many ways to conduct a watershed assessment and many data forms that have been developed, ranging from a simple, one-page form, to more complex forms that measure a wide variety of parameters. Project and watershed assessment approaches suitable for local conservation groups are available from the Natural Resources Conservation Service at the web site: *www.wcc.nrcs.usda.gov/ watershed/products.html.*

Another general assessment that examines several factors is the Rapid Stream Assessment Technique (RSAT) developed by the Metropolitan Washington Council of Governments (Washington, DC). The RSAT combines elements from several protocols and leads survey teams to assign scores to the stream's channel stability, erosion and deposition, in-stream habitat, water quality, riparian habitat, and biological indicators. RSAT can be downloaded from the Internet at *www.stormwatercenter.net/ intro_monitor.htm.*

A watershed assessment developed by Dale Pfankuch in 1978, *Stream Reach Inventory and Channel Stability Evalua-tion,* looks at the stability of upper and lower stream banks; the size and composition of materials on the stream bottom; the width, depth, and velocity of the stream; and other factors (see Appendix C). Volunteers may need technical help to perform Pfankuch's assessment method because it does require some ability to recognize stream corridor characteristics, such as bankfull discharge. Local and state environmental departments may also have regionally tailored protocols.

Another option is to work with the project's technical team to develop the most appropriate type of watershed assessment form based on project goals and the time and expertise of the group. For example, a watershed assessment that will determine priority sites and/ or reference reaches for stream bank enhancement projects should concentrate primarily on the banks, although other stream corridor components should also be considered. Likewise, a stream enhancement project that focuses on riparian buffer improvements should examine soil qualities and the relative abundance of native vegetation and invasive plant species.

Many of the questions used on an assessment form are subjective and open to interpretation. It is important to ensure that the data collected is consistent. To improve consistency among volunteers, discuss with the group what scores to assign different stream conditions. If appropriate, take photographs that indicate certain features that need to be present in order to obtain a certain score. This will help control consistency of the data if the group splits into smaller teams to assess stream reaches independently. The watershed assessment scores can be used to prioritize sites for enhancement. A good way to organize this information is to indicate stream segments of different quality on the map using different colored markers to represent each assessment score (poor, fair, good, excellent, etc.) (Figure 26).

Review the "Watershed" section in Chapter One for more information on finding watershed boundaries.

More information on the concept of bankfull is included in Chapter One, while information on methods of field determination are found in Chapter Three.

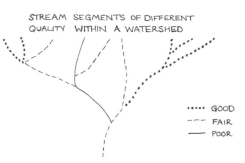

STREAM SEGMENTS OF DIFFERENT
QUALITY WITHIN A WATERSHED

⋯⋯ GOOD
--- FAIR
— POOR

Figure 26. Simple line graphics can illustrate stream segments of different quality within a watershed. (Izaak Walton League of America figure)

Prioritize Sites for Enhancement

After collecting data on sites throughout the watershed, use the watershed map to mark segments that are high, medium, and low priorities for enhancement. The best course of action is to stabilize sites located higher or upstream in the watershed before those downstream. A stabilized site downstream in the watershed may quickly become degraded again because of problems caused by the upstream site.

Overall, the protection of existing, good quality stream sites is the soundest approach to watershed enhancement. In other words, if there are reaches of a stream with healthy riparian vegetation and good habitat for aquatic species, first consider ways to protect these areas from degradation. A buffer of continuous intact riparian vegetation is more effective at protecting streams than fragments of stabilized stream banks. Enhancement sites that connect intact riparian areas should be considered a high priority.

In some cases, groups may want to choose a project site because it is convenient or highly visible such as a local school or community center. If the primary goal is community education rather than improving water quality, such a site may be appropriate. However, unless designed in the context of an overall watershed assessment and the resulting enhancement plan, such sites should only be used for stream buffer planting projects. This will prevent wasting funds and other resources on a more involved enhancement project that may not provide much benefit for water quality. More importantly, limiting the project to a buffer planting will decrease the chances that the project could cause damage downstream or be washed away due to upstream conditions. Groups planning site-specific projects should still consider conducting a limited watershed assessment. Every enhancement project requires some background research to make sure the project will fit into the community's future plans for the area.

Other important factors to consider when planning volunteer projects are safety and permission. For example, a site along a high traffic road with no sidewalks is not appropriate for volunteers. For additional information on safety, see Appendix D. Always contact local landowners before monitoring to make sure you are not trespassing. Ask for permission if it's necessary to cross private land. Most landowners will give permission for the survey and they might even want to help or know the survey results.

Consider Whether to Use Stream Classification

Stream classification is a tool used by many stream restoration professionals to identify patterns in the natural form and dynamics of stream channels. Classification allows streams to be placed into different categories based on shared characteristics such as location in the landscape, the shape of the stream channel, and the size of sediment carried by the stream.

In addition to helping identify appropriate reference reaches for a stream segment, placing streams into categories provides a succinct way of communicating

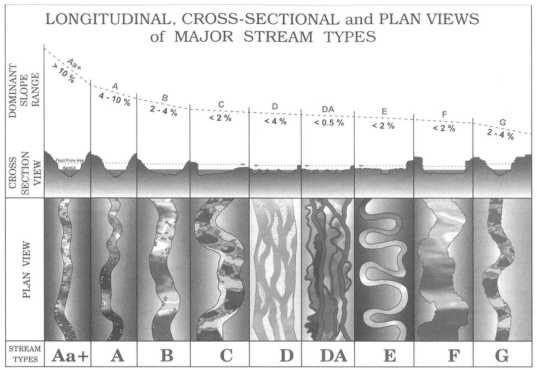

Figure 27. Rosgen's classification system provides a useful methodology for grouping common stream types. (Rosgen, 1996)

a large amount of information. Classification can also help stream professionals, such as hydrologists or engineers, to select enhancement measures for specific segments because streams that share certain characteristics tend to behave similarly.

There are stream classification systems used by professionals that might be helpful in determining the scope of a project for volunteer groups. Some of the most common systems include those developed by Schumm (1977), Montgomery and Buffington (1993), and Rosgen (1996). Of these, one of the most detailed and widely used is the classification system developed by Dave Rosgen. Rosgen's stream classification system is based on field measurements of hundreds of natural, stable stream channels. These channels were then grouped into categories based on common morphological patterns. The first level of Rosgen's four levels of classification is based on channel slope, channel shape, and channel patterns

(single, multiple, braided) (Figure 27). The second level looks more specifically at some measurable characteristics of stream morphology. It is important to note that different classification systems are designed for use in different parts of the country.

More information on these systems can be obtained through the listings in the Bibliography at the end of the book and in the Watershed Stewardship Resources document available on-line at *www.iwla.org/sos/resources.*

After a watershed assessment has been used to identify enhancement priorities, reference reaches, and other background information, a more detailed site inventory is needed for each project site. Chapter Three describes various site inventory methods that may be useful for enhancement projects with different goals.

Although volunteers might be tempted to try to classify a stream using data forms from one of the systems mentioned in this chapter, accurate classification relies on mastery of some very technical skills.

Chapter Three

Performing a Site Inventory

*A*fter conducting a watershed assessment (or working with local government agency staff to develop a project consistent with a previously approved watershed assessment) and selecting a site for a stream enhancement project, additional information should be gathered through a site inventory.

A site inventory is the next step in preparing for a stream enhancement project. It is more involved and in-depth than a watershed assessment. It determines the types of enhancement projects needed and establishes baseline conditions for future monitoring (Figure 28). The information that will be gathered will depend greatly on the project's goals (see Appendix E).

This chapter provides information relevant to various stream enhancement projects with different goals.

The site inventory information required for a particular project depends on the project's goals. Therefore, all the data measurements described might not be necessary for every stream enhancement project.

Several of these measurements also require advice from the project's technical team. The descriptions of data provided

Figure 28. Site inventories establish baseline information needed for enhancement project design. (Izaak Walton League of America photograph)

are intended to shed light on the possible techniques involved. Seek additional input from local sources when performing data collection and obtain permission from landowners before entering private property to evaluate a stream.

Choose Site Inventory Parameters

The site inventory should focus on data directly related to the goals of the project. For example, a stream bank enhancement project will need data on the bank stability, streambed and bank materials, base flows and storm flows, and buffer vegetation. There are many site characteristics to consider. Consult the technical team to develop a site inventory with parameters that are appropriate and significant to your watershed and project.

One ecological inventory that might be particularly useful to volunteer groups is the Stream Visual Assessment Protocol developed by the Natural Resources Conservation Service for use by conservationists with little biological or hydrological training. This protocol allows individuals to rate the health of a site based on up to 15 elements, which include bank stability, channel condition, riparian vegetation, canopy cover, in-stream fish cover, and barriers to fish movement. The Stream Visual Assessment Protocol is available on the Internet at: *www.wcc.nrcs.usda.gov/water/quality/ frame/wqam/Guidance_Documents/ guidance_documents.html.*

Another resource to consult when investigating site conditions is the Farm Assessment System/Home Assessment System *(www.uwex.edu/farmasyst/, www.uwex.edu/homeasyst,* or *www.uwex.edu/AgMES)* program. This program has developed voluntary, confi-

dential environmental assessments for farms, homes, and rural properties. Participants complete worksheets to evaluate risks to their family's health, the environment, and offers suggestions to consider for better farmland management practices.

The following pages provide brief descriptions of different attributes that can be used to describe the stream corridor condition during a site inventory and how that information can influence project design. Each subsection in this chapter focuses on a specific site inventory parameter. For each parameter, the following information is provided: 1) introduction to the parameter; 2) how to collect data; and 3) how volunteers can participate in data collection. Some of the site inventory measurements require expert assistance and the most important volunteer action could be requesting stream data from agencies and helping to organize the activity. Consult the members of the technical team to discuss specific volunteer duties.

Develop a Site Map

A site map, drawn during the site inventory, depicts detailed information about the stream corridor and its features not available on printed maps. The site map serves as a visual depiction of the stream that, along with other information collected during the site inventory, can be used to determine appropriate enhancement techniques. The site map also serves as a base map for project plans.

First, draw the stream and its floodplain. Include the shape of the stream channel, in-stream characteristics such as point bars and riffles, areas of existing riparian vegetation, adjacent land uses,

Figure 29. When it rains, a large amount of water runs off impervious surfaces, (A) enters the storm drain system, and (B) is directly discharged into a stream. (Izaak Walton League of America photographs)

property lines, **storm drain outfalls**, and any structures such as bridges (Figure 29). Make copies of the map to record additional information during the site inventory, such as location of underground utility lines and debris jams in the channel. Measure the length of the stream segment to be enhanced and mark this on the map.

Volunteers can draw accurate and useful site maps. Although drawing a map might seem difficult at first, it is very useful for all people involved in the project to try this activity. Drawing site maps will help volunteers analyze site features carefully and interpret enhancement plan drawings for other project sites (Figure 30).

Figure 30. A site map of a stream created during a site inventory. (Harrelson and Potyondy, 1994)

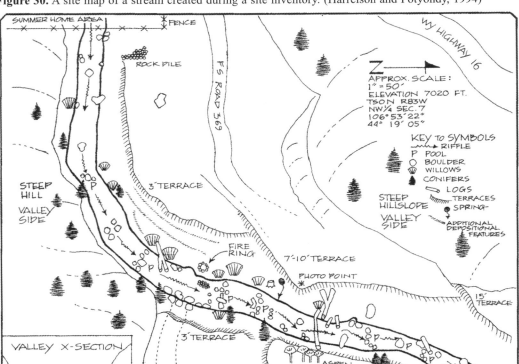

Figure 31. Severe erosion along this stream necessitated corrective action. These photographs clearly show (A) the problem and (B) the work that was done to correct it. (Licking County [Ohio] Soil and Water Conservation District photographs)

Establish Photo Points

Another general inventory and monitoring tool is to make a photographic record of the stream at different times of the year and during periods of high, low, and normal water flow. This visual record can be very helpful in recording characteristics of the stream and changes in the stream's shape and size over time. Photographs are powerful tools that can be used in public meetings and other forums to explain and demonstrate stream alterations and potential problems (Figure 31).

Photo points are reference spots along the stream. It is important to take photographs from the same point and at the same angle *each time* so they can be compared directly to one another. Photo points can be marked at the site with permanent stakes or can be established at landmarks such as trees or **culverts**. Photo points should be marked and numbered on the site map. Include a person or yardstick in the foreground to show scale. Keep a written record of photographs as they are taken, and record the flow level of the stream at the time the picture was taken if possible. When the film is developed, mark each photograph with the date, the photo point number, and the water level of the stream. Videotaping the site also is recommended. Photographs should be taken at least once each season and during high flows. To determine water level of the stream accurately, a stream gauge may need to be installed. More information about installation of stream gauges follows in the next section.

With some guidance, volunteers should be able to establish photo point locations, install permanent stakes to mark these locations, take photographs, and record information.

Measure Stream Flow

Stream flow or discharge (the words flow and discharge are used interchangeably throughout this document) recorded in cubic feet per second (cfs), is a measurement of the volume and speed of water

moving through the stream channel. It is important to understand the characteristics of stream flow when enhancing stream banks because the project design will need to accommodate the water that typically flows through the stream as well as floods and droughts.

The most basic information about the water in the stream includes whether the stream is perennial (water always flows in the stream), intermittent (water flows seasonally), or ephemeral (water flows during and immediately after rain events). It also is important to know frequency, duration, and water levels of base flow and flood events. This information shows how frequently stream banks and riparian corridor vegetation will be flooded and will help with the selection of appropriate plant materials for the site. Stormwater flows are problematic to streams in urban areas and can destroy poorly planned stream enhancement efforts. Stream flow is less significant to enhancement projects that are limited to planting trees and shrubs in riparian zones.

In order to accomplish many types of enhancement projects, it is important to understand bankfull discharge. Bankfull discharge is the flow that is most effective in shaping and maintaining the natural stream channel. Knowing the bankfull discharge is essential to the design of stream enhancement projects that include channel reconstruction, which involves changing the shape of the stream channel using earth-moving equipment. In addition, projects that involve work below the ordinary high water mark, a legal designation which is sometimes equated with the bankfull stage, usually require more strict permits than those for work above this mark. Therefore, recognizing the bankfull stage of the stream is important in deter-

mining which permits may be needed to complete enhancement projects.

The easiest way to obtain information on the stream flow is to check with your state department of transportation. Also, if any other state or local agencies have worked on the stream, they may have stream flow data available. Another good source of information on stream flow is stream gauge data. The US Geological Survey maintains gauging stations that measure water levels and flow along many streams across the country. Contact the local office of the Geological Survey or visit their web site at *http:// water.usgs.gov/* and click on "USGS Information by State" to find out if stream information is being recorded near the project site.

Determining bankfull discharge requires a determination of the bankfull stage, a measurement of the cross-sectional area of the channel at the bankfull stage, and a measurement of the velocity of water when the stream is at the bankfull stage. All of this information is necessary because bankfull discharge equals the cross-sectional area at the bankfull stage multiplied by velocity. In theory, the bankfull stage of a stream can be measured in the field. However, the bankfull stage is not always easy to determine, particularly in urban streams that may be out of equilibrium because of altered upstream land use. It is also important to distinguish whether the stream bank features used to indicate the bankfull stage are present because of the current regime of water flow or are a result of an earlier, different stream flow.

There are some methods of estimating the bankfull discharge or flow that can

For background information on stormwater and bankfull discharge, please refer to Chapter One.

Figure 32. A change in the particle size of bank material, such as the boundary between coarse cobble or gravel with fine-grained sand or silt, may indicate bankfull level. (Harrelson and Potyondy, 1994)

be used in place of field measurements. If the stream has a Geological Survey gauge station, the bankfull discharge can be estimated by examining the flows recorded over time. On average, the bankfull discharge is associated with floods that occur with a frequency equal to about once every 1.5 years. If streams do not have stream gauge data, estimates of flow duration and the frequency of extremely high and low water flows may be based on regional flooding statistics throughout the region. State transportation departments often apply equations based on the relationships between physiological characteristics of streams with Geological Survey gauging stations to size culverts and rural road bridge openings on streams without gauge stations in the same region. In order to use these equations, the user must first determine if the project site has similar characteristics to the gauged site, including annual precipitation, drainage area, land uses, soil and bedrock characteristics,

stream slope, and water storage available in lakes and wetlands.

Another option to consider if the stream does not have gauge data is to install a gauge at the stream site. A gauge can be as simple as a metal yardstick attached to a fence post. Install this gauge during low water flow at the water's edge. Data need to be collected at regular intervals such as once a day or once a week, as well as during storms, to provide information on a variety of changes in water levels.

Although gauge station data may be used to estimate the bankfull stage of a stream without field measurements, field measurements are preferable because they consider local physical surroundings. Field indicators of bankfull stage are often difficult to determine and require the expertise of trained and experienced personnel. Seek out members of the technical team to help with field measurements.

The height of depositional features, such as the tops of point bars (areas of sand and gravel typically found along the inside of meander bends), is one of the best indicators of bankfull stage. Also look for other areas of deposition, typically consisting of fine sands, that tend to occur along many channel types (Figure 32). A change in vegetation may serve as an indicator in the western United States. However, in the eastern part of the country, a change in vegetation near stream banks tends to be a poor indicator of bankfull because vegetation often grows below bankfull elevation. The following picture provides an overview of where bankfull stage might be located in a variety of streams.

Figure 33. The lines drawn across this picture of a stream indicate the stream's bankfull level. In the field, it can be useful to mark the bankfull level with flags to get a good visual display of the bankfull line. (USDA Forest Service, 2003)

Pin-flags can be inserted along the bank where field indicators of bankfull stage are located to help get a better visual depiction of the bankfull line. Photographs can be taken of the bankfull stage marked with the flags (Figure 33). More information on field indicators of bankfull stage can be found in the following Forest Service materials: *Identifying Bankfull Stage in Forested Streams in the Eastern United States* (available as a VHS or DVD), *Guide to the Identification of Bankfull Stage in the Western United States* (available as a DVD or CD-ROM) or *Stream Channel Reference Sites: An Illustrated Guide to Field Technique* (available in hard copy). Resource information is listed in the Bibliography of this book and the Watershed Stewardship Resources document available on-line at *www.iwla.org/sos/resources.* Although these publications can provide assistance with field determination, it is best to work with the technical team to determine the elevation of bankfull stage.

In order to convert water levels into discharge, measurements of the cross-section and the velocity at the gauge site are needed. Velocity can be measured using a velocity meter or through a formula that requires additional measurements. More information on field measurements of channel cross-section is provided later in this chapter in the section "Investigate Stream Channel Characteristics" (pp. 41–45). Methods and calculations for converting the stream channel measurements into the cross-sectional area and for determining velocity can be found in the Forest Service manual *Stream Channel Reference Sites: An Illustrated Guide to Field Technique,* listed in the Bibliography. Once the bankfull stage is determined and measurements of the channel cross-section have been taken, this information can be translated into the bankfull discharge using the equation

$$\text{Discharge} = \text{velocity} \times \text{cross-sectional area.}$$

When field determinations of bankfull discharge are needed, volunteers should enlist the project's technical team. Volunteers can obtain and analyze stream gauge data and speak with local and state agency staff to find existing information on stream flow. Technical team members should also assist with construction and installation of stream gauge stations and with calculations of cross-sectional area, velocity, and discharge.

Figure 34. Consult experts for help interpreting soil survey maps. (Izaak Walton League of America photograph)

Assess Stream Corridor Soils and Erosion

The primary function of most stream bank enhancement projects is to slow bank erosion. Understanding the type of soil found at the site and its characteristics will help determine the best methods to achieve bank stabilization. Some soils are resistant to erosion while others erode easily. Clays tend to hold water, yet sandy soils drain easily. The soils present at a stream site — along with streamside vegetation — affect infiltration, groundwater storage, runoff, and slope stability. The vegetation chosen for the enhancement project also depends on soil types because some plants require particular soil types to thrive.

The enhancement techniques chosen also will depend upon the severity of erosion. In healthy, stable streams, banks are covered with a mix of vegetation that helps reduce erosion, provides shade, and offers food for aquatic wildlife. Some erosion of the banks is normal in stable streams and is balanced by deposition of sediment elsewhere. Rapid, severe erosion, however, degrades water quality and creates additional bank instability downstream.

Stream Bank Soils

The Natural Resources Conservation Service publishes soil surveys for every county (see Watershed Stewardship Resources available on-line at *www.iwla.org/sos/resources*). Soil maps may also be available from the Forest Service or Bureau of Land Management. Soil surveys include maps that show soil types, as well as detailed descriptions of the characteristics of each soil type. Characteristics described in soil surveys include soil texture, whether the soil is well or poorly drained, appropriate land uses for the soil, and other information (Figure 34). While soil surveys provide helpful information, it is also important to examine soils in the field because the soil maps are developed based on aerial photographs, so boundaries between soil types are sometimes inaccurate.

Another reason that soil in the field might not match the soil listed in the soil survey is that the soil may be **fill**, especially along streams that have been altered unnaturally. Fill is soil that has been dumped to fill in low spots, close off meanders, or create more area for land use. Fill soil along a waterway may be very unstable, nutrient poor, and/or contaminated.

Field examination of the soil can include digging a soil pit; examining the different soil layers, textures, and colors at various depths; and taking soil samples for testing. State Cooperative Extension offices may be able to test soil samples collected at the enhancement site. Testing can reveal the pH, nutrients, texture, and other soil characteristics that can affect the

types of plants that will thrive on the site. Contact local Natural Resources Conservation Service offices for assistance with examining soil characteristics in the field (Figure 35).

Stream Bank Erosion

A preliminary field evaluation of stream bank erosion can be similar to that performed during the watershed assessment. A walk along a stream site can assist in determining whether most of the banks are exposed or vegetated. Sketch the stream and mark areas of severe stream bank erosion or exposed soil. Note whether there are areas of the bank that are caving in or larger sections of land that are sliding into the stream. Identify any artificial structures along the banks such as concrete, rock **riprap**, metal sheet piling, or other stabilizing structures (Figure 36).

Figure 35. It is important to verify soil map data by checking soil characteristics in the field. (Izaak Walton League of America photograph)

The next step will be to record the slope and height of the banks from the water's surface at normal flow. Slope is most often expressed as a measurement of the change in vertical distance over (i.e., divided by) the change in horizontal distance — or, as "rise over run." A slope that has one foot of vertical change (rise) for every two feet of horizontal change

Figure 36. A sketch of stream bank erosion and deposition sites can serve as a useful guide to stream enhancement. (Izaak Walton League of America figure)

EROSION

DEPOSITION

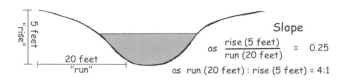

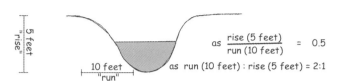

Slope

as $\dfrac{\text{rise (5 feet)}}{\text{run (20 feet)}}$ = 0.25

as run (20 feet) : rise (5 feet) = 4:1

as $\dfrac{\text{rise (5 feet)}}{\text{run (10 feet)}}$ = 0.5

as run (10 feet) : rise (5 feet) = 2:1

Figure 37. Slope can be quantified as the relationship between change of horizontal distance and change of vertical distance. This diagram illustrates two methods of quantification — "rise over run" and the ratio of run to rise. (Izaak Walton League of America figure)

(run), for example, would have a slope of ½ (rise over run), or 0.5. Another way of expressing slope, one often used by engineers, is as a ratio of the change in horizontal distance to the change in vertical distance (Figure 37). Using this method, the example of one foot of vertical change for every two feet of horizontal change would be expressed as 2:1. A slope expressed as 0.5 by the former method would be identical to a slope of 2:1 as expressed by the latter method. Steeper slopes are less stable and more susceptible to erosion than are more gradual slopes. Slopes no greater than 0.33 (3:1) or 0.25 (4:1) are preferred for vegetation to grow on the banks.

Enhancement projects often require re-grading of the banks to restore a stable, gradual slope. To measure the bank height — vertical distance — use a surveyor's rod or other tall measuring tool. Bank height should be measured from the surface of the water at normal flow to the top of the bank. To determine bank slope, measure the horizontal distance between the surface of the water and the top of the bank. The slope is equal to the vertical distance divided by the horizontal distance.

In addition to examining the current stream conditions, erosion can be determined by measuring changes in stream bank location. Aerial photographs may reveal the amount of erosion that has occurred over time, but they may not exist for the site or may be too old to provide an accurate estimate of recent changes in overall channel shape. Aerial photographs can be obtained from federal, state, or local government agencies (e.g., farm service agencies, county planners, etc.). Photo points can provide a good visual depiction of changes in channel shape, including the progression of erosion. Measurements of the channel cross-section (see pp. 43–44) taken at the same location over time also may reveal changes in stream channel shape from erosion and deposition.

Another method of measuring erosion over time is to use bed and bank pins. The pins can be made by using any durable, straight rod inserted horizontally at regular intervals into a stream bank, leaving a standard length exposed. Rebar stakes, which can be purchased at most hardware stores, work best. Several rebar stakes are driven into different points across the streambed and at different heights along the bank at a particular cross-section of the stream. Bed and bank pins are then checked at regular time intervals. The change in the exposed length of the pins is used to determine the rates of erosion and deposition (Figure 38). Photographs of bed and bank pins provide evidence of these

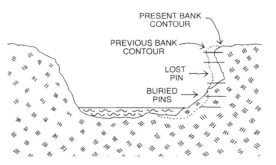

Figure 38. Stream bank erosion pins can be used to measure rates of erosion and deposition over time. (Harrelson and Potyondy, 1994)

changes over time. More information on installing bed and bank pins, as well as information on scour chains (a similar method of measuring erosion and deposition), can be found in the Forest Service manual *Stream Channel Reference Sites* (see Bibliography).

Enhancement projects that aim to curb stream bank erosion should address areas with a high potential for future erosion. Field measurements can help determine the potential for future erosion.

The **bank erosion potential rating** is a measurement that indicates the ability of stream banks to resist erosion based on several factors, such as the composition of the stream bank materials, the bank slope, and the root density of riparian vegetation. Other measurements that provide clues about potential future erosion are the **width-to-depth ratio** and the **entrenchment ratio**.

The width-to-depth ratio is a measurement of the average bankfull width over the average bankfull depth. As width-to-depth ratios increase, the channel becomes wider and shallower, forcing the water to spread out, thus leading to more bank erosion. The entrenchment ratio is used to describe the degree of vertical containment of a stream channel (Figure 39). The banks of an entrenched channel (Figure 40) are high enough that floodwaters cannot overtop the banks. During

Figure 39. Various types of streams are characterized by different entrenchment ratios, which can be determined by dividing flood-prone width by bankfull width. (Rosgen, 1996)

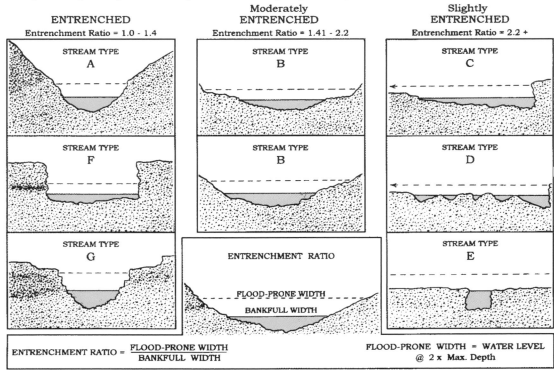

Figure 40. Increased runoff during decades of grazing has contributed to this stream becoming entrenched. (Jerry N. McDonald photograph)

floods, the water hits the banks with a greater force and causes erosion. Changes in these ratios over time indicate how a stream section is responding to surrounding changes in land use. Additional information on width-to-depth ratios can be found in Rosgen's *Applied River Morphology* (See Bibliography).

Volunteers can obtain information on soils through the soil survey and can collect soil samples for testing. More detailed information on soils can be collected in the field with the help of Natural Resources Conservation Service staff or the project's technical team. With some training, volunteers may be able to perform the preliminary field evaluation of erosion, determine the bank slope and height, and install and use bed and bank pins. The project's technical team should be enlisted when examining bank erosion potential ratings, width-to-depth ratios, and entrenchment ratios.

Investigate Stream Channel Characteristics

Stream channel characteristics are important for any stabilization project because the bank needs to be able to handle the volume of water both during ordinary flow and extreme flood events. Also, changes in the shape of the channel recorded over time can indicate areas of erosion and deposition, so baseline information on channel features are necessary for comparisons. Important dimensions of the channel to measure include the stream's dimension, or cross-section, which is perpendicular to the water flow, and its longitudinal profile, which shows the change in slope from upstream to downstream. The dimensions of the stream's cross-section provide clues about whether the stream is connected properly to its floodplain. Cross-section and profile surveys taken over time at the same location provide information on how the channel shape is changing in response to changes in surrounding land uses.

Measurements of the channel characteristics such as stream dimension, profile, and pattern over time can help determine whether the stream can recover on its own or, if not, the extent of active stabilization that is needed. If the site inventory reveals a need for channel reconstruction, stream professionals should determine the channel size and shape needed to withstand the flows and other stresses on the stream. Again, it is important to examine a stream in the context of local and regional conditions. A site that might be classified as an unstable stream in the Rocky Mountain foothills might be considered stable in the Mississippi Delta. The next section will highlight data collection techniques for stream channel features and streambed characteristics.

Stream Channel Features

Many important stream channel characteristics can be represented on the site map. Some of the characteristics that should be depicted include the direction of water flow, the stream pattern (meander, braided, or straight), and the channel width. Also, include sand or gravel bars, pools, riffles, areas of woody debris, large boulders, and any dams or other obstacles to fish passage (Figure 41).

> Chapter One describes the three different views of the stream corridor: the stream dimension, the stream pattern, and the stream profile.

Streambed Characteristics

The composition and characteristics of the streambed influence channel shape, erosion rates, and the supply of sediment to other parts of the stream, and they dictate the way water moves through the system. Streambed characteristics include the type of **substrate**, distribution of particle size, and areas of scour and deposition. The substrate may be bedrock, boulders, cobbles, gravel, sand, silt, or some combination of these (Figure 42). Check with members of the project's technical team to determine the amount of information on streambed characteristics needed for the project's goals. Professionals can also help to gather streambed information.

Visual observations of the streambed can be deceiving. Although boulders and cobbles may seem dominant, there may be more sand particles that fill in the gaps around these larger pieces. One way to get a good estimate of substrate particle size is to do a pebble count. This involves walking across the stream and stopping at regular intervals, then picking up the particle closest to one's toe and measuring it. After gathering 100 samples, the average particle size can be determined. More information on pebble counts can be found in the Forest Service manual *Stream Channel Reference Sites* and in Rosgen's *Applied River Morphology* (see Bibliography).

Areas of scour and deposition include natural pools and sandbars. Sandbars that develop in the middle of the stream may indicate that there is too much sediment moving through the system due to erosion upstream. Island sandbars also may show that the stream has widened to the point that flow slows down and sediment drops out in the middle of the channel rather than on the inside of meander bends. Scour is natural downstream of riffles or large

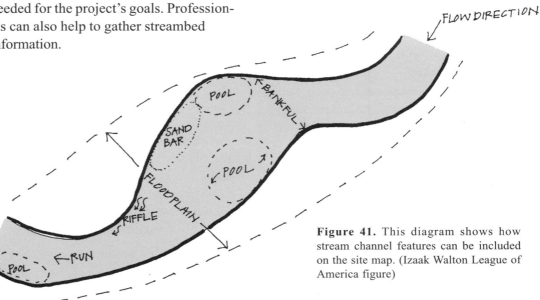

Figure 41. This diagram shows how stream channel features can be included on the site map. (Izaak Walton League of America figure)

A Handbook for Stream Enhancement & Stewardship

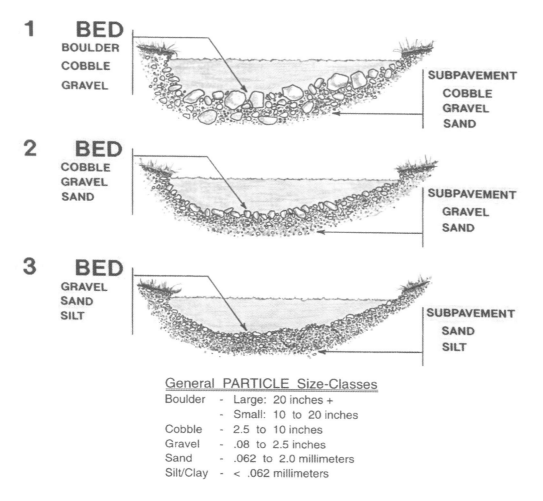

1 BED BOULDER COBBLE GRAVEL

SUBPAVEMENT COBBLE GRAVEL SAND

2 BED COBBLE GRAVEL SAND

SUBPAVEMENT GRAVEL SAND

3 BED GRAVEL SAND SILT

SUBPAVEMENT SAND SILT

General PARTICLE Size-Classes

Boulder - Large: 20 inches +
- Small: 10 to 20 inches
Cobble - 2.5 to 10 inches
Gravel - .08 to 2.5 inches
Sand - .062 to 2.0 millimeters
Silt/Clay - < .062 millimeters

Figure 42. These stream cross-sectional diagrams demonstrate the general categories of channel bed and subpavement materials. (Rosgen, 1996)

rocks and produces deeper pool habitats. There may also be areas of excessive scour downstream of a debris jam or an improperly installed in-stream enhancement technique. All areas of natural and excessive scour and deposition should be recorded on the site map to help prioritize sections of the stream bank for enhancement.

Cross-Section

Changes in the channel dimension display how the stream is responding to changes in the surrounding landscape. The channel cross-section is a slice of the stream channel shape that is perpendicular to the water flow and illustrates the channel width and depth. To measure the cross-section, locate a straight reach of the stream, not at a meander bend or riffle.

Drive a piece of rebar into the ground on each side of the stream a few feet back from the top of the bank. Stretch a nylon line and a measuring tape between the rebar stakes and secure them. Use a line level to ensure that the line is straight. Measure vertical distance along the horizontal line using a surveyor's rod. Take measurements every one or two feet. Take additional measurements at certain features, such as at the top of the bank, bankfull stage, any breaks in slope along the bank, and the deepest point of the

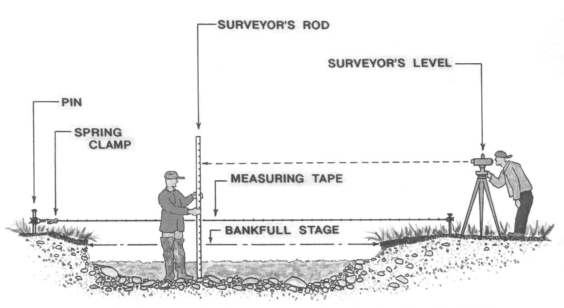

Figure 43. By measuring the cross-section of a stream, one can determine the width and depth of the stream channel, as well as certain key features. (Rosgen, 1996)

channel. Record each horizontal and vertical distance, and note specific features (Figure 43).

To draw the cross-section, convert the differences in elevation between the nylon line and the ground into actual elevations by subtracting each measurement from the true elevation above sea level of the nylon line. If the true elevation is unavailable, use an estimate of this height. By plotting the distance on the horizontal axis and the elevation on the vertical axis, an accurate representation of the channel cross-section can be drawn. See Appendix F for a data sheet upon which to record cross-section data. This cross-section drawing can then be used as a base to map out potential stabilization techniques and as a tool to track changes in the stream channel's shape over time.

Longitudinal Profile

The longitudinal profile shows the change in slope along the stream segment from the point farthest upstream to the point farthest downstream. Often, the slope is estimated using aerial photographs and topographic maps. The valley slope can be determined using the elevation information on topographic maps. The sinuosity, or curviness, of the channel can be determined with aerial photographs. Valley slope over sinuosity provides a generalized estimate of channel slope. More accurate measurements of channel slope can be taken in the field using surveying equipment. Measurements of elevation of the lowest point in the channel are taken at several intervals along the stream site. The longitudinal profile is most often used to help determine a stream's classification, which can in turn provide some insight about appropriate enhancement techniques.

Volunteers can record stream features on the site map. With proper training, volunteers will be able to take measurements of bed materials using a pebble count, if determined necessary by the technical team. With the proper equipment, volunteers also can take channel cross-section measurements (Figure 44). Technical team members can assist in determining locations for cross-sections. Technical

Figure 44. Periodically calculating the cross section of their stream channel allows these volunteers to track changes in channel shape over time. (Izaak Walton League of America photograph)

team members, or others experienced in the use of surveying equipment, should assist in measuring the stream's longitudinal profile, if this measurement is necessary for the project.

Evaluate Riparian Vegetation

The width and quality of the vegetated buffer between a stream and surrounding land contributes to the wildlife habitat, water quality, and stability of the stream. The wider and more diverse the vegetated buffer, the greater benefit it can provide for the stream and its inhabitants. Many states recommend a minimum buffer of 50 to 100 feet to accommodate significant ecosystem functions. Community stream bank projects can include establishing or expanding buffer vegetation along a stream. Also, a catalog of the existing plants can provide clues to selecting

appropriate vegetation for enhancement. Information on the soils and flooding regime gathered as part of the site inventory can help with plant selection. A plant inventory may also reveal a need to remove invasive species.

An inventory of riparian vegetation should take place during the growing season when plants can be most easily identified. A good place to start is measuring the width of the buffer in several places using a measuring tape and recording the average.

Next, determine the percentage of banks currently covered by trees, shrubs, grasses, or exposed soil. This information helps to identify enhancement opportunities. The area covered by the tree canopy is important because canopy cover provides the shade required for aquatic life. Vegetation along the banks also prevents erosion because plant roots hold soil in place. Streams with vertical, eroded banks often are good candidates for vegetative enhancement techniques that involve grading the banks to a gentler slope and planting vegetation. However, if the canopies of large trees growing adjacent to the stream shade the banks, it might be difficult for vegetation to grow. Investigate using native plants that grow in shaded areas.

Estimates of canopy cover indicate the quality of the habitat provided by the riparian vegetation. The vegetative canopy provides food and shelter for wildlife, as well as cooler water temperature, thus more dissolved oxygen for aquatic organisms. Canopy cover estimates can be made

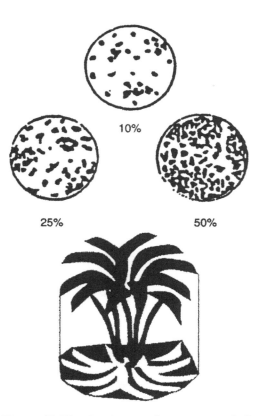

10%

25%   50%

Figure 45. Visual estimates of canopy cover indicate the quality of habitat provided by riparian vegetation. The circles represent what a person would see when standing along the streambank and looking up at the canopy. (Izaak Walton League of America figure)

visually, similar to estimates of percent cover on the banks (Figure 45). There are also methods that use equipment, such as a **spherical densiometer**, to aid in estimating the percentage of canopy cover on stream banks.

A spherical densiometer is a concave mirror with a grid of equally spaced dots or lines on the surface. The mirror reflects vegetation overhead. To use a spherical densiometer to calculate a percentage of canopy cover, divide the number of dots covered by the reflection of the canopy vegetation by the total number of dots. Measurements should be taken at regular intervals, such as every two feet, while walking across the stream from one bank to the other. Another method to estimate canopy cover is to use an ocular tube

constructed from PVC pipe or other rigid tubing with cross hairs (made with thread or dental floss) on one end (Figure 46). Walking in straight lines across the stream in several places, stop every few feet, and look up through the tube. For each stop, record in a field notebook when vegetation is present or not present in the cross hairs. At least 50 samples should be taken to determine the percent canopy cover using this method. The percent canopy cover can be determined by dividing the number of samples taken where vegetation was present by the total number of samples taken.

Another important component of the riparian vegetation inventory is to identify the species of plants within the riparian zone (Figure 47). Native trees, shrubs, and grasses with deep root structures are most capable of holding stream banks in place. This information should be recorded both as a species list and on the site map to show species location. Note any threatened or endangered species and any exotic or invasive species. If invasive plant species are present on the site, their removal needs to be a component of the stream enhancement project.

With the proper training and tools, volunteers may be able to conduct all of the field measurements required for surveying riparian vegetation and canopy cover. Technical team members, or others with experience in plant identification, should assist in a species inventory.

Figure 46. This ocular tube can be made inexpensively by using PVC pipe and dental floss. (Izaak Walton League of America figure)

Figure 47. Plant field guides and identification keys are helpful tools when investigating local flora. (Izaak Walton League of America photograph)

Examine Water Quality

Water quality is an important factor to consider when planning any stream enhancement project. Understanding the quality of the water will help to establish effective goals. Simple physical indicators, such as color and odor, can provide clues to possible sources of pollution (Table 1). Further chemical or biological tests provide a more accurate picture of stream water quality and biological diversity. See the Watershed Stewardship Resources available on-line at *www.iwla.org/sos/ resources* for sources of chemical and **biological monitoring** equipment.

Sensory Assessment

Walk the stream segment and take note of any unusual stream colors and odors. Do not rush to conclusions, but investigate potential causes further by examining the land uses within the watershed. As the chart on the next page indicates, natural processes can cause many unusual colors and odors. Document the presence of unusual colors and odors by marking their location on the site map and with photographs.

Biological Indicators

The wildlife that lives in and along waterways is an excellent source of information about water quality because all organisms require specific conditions to live. Many of the insects, crustaceans, fish, and other organisms differ in their sensitivity to water pollution. The presence and quantity of pollution-sensitive organisms provides a reliable measurement of water quality.

Macroinvertebrates

The Izaak Walton League's Save Our Streams (SOS) program uses the presence of **benthic macroinvertebrates** to measure water quality. **Macroinvertebrates** are aquatic animals that are large enough to see with the naked eye (macro) and have no backbone (invertebrate). Benthic macroinvertebrates live in the stream bottom, and include insect larvae, adult insects, and crustaceans.

Stream-bottom macroinvertebrates are good indicators of water quality because they differ in their sensitivity to water pollution (Figure 48). Some benthic

Figure 48. Stonefly larvae (left) are very sensitive to water quality and cannot survive in degraded waters. Lunged snails (right), on the other hand, are tolerant organisms that can persist in areas of degraded water quality. (Izaak Walton League of America figures)

macroinvertebrates are very sensitive to pollution and cannot survive in polluted water. Others are less sensitive to pollution and can be found even in very polluted streams. Benthic macroinvertebrates usually live in the same area of a stream for most of their lives. If the water quality is generally poor, or if a pollution event occurred within the past several months, it will be reflected in the macroinvertebrate population.

Table 1. Common Water Quality Indicators and Their Potential Causes.

COLOR	POTENTIAL CAUSE
Green water	May be caused by excessive algae growth, or may indicate nutrient enrichment from sewage, fertilizers, or livestock waste runoff.
Orange-red water	May be caused by acid mine drainage, oil runoff, or industry discharges. May be natural in areas with iron-rich water.
Light brown, opaque water	May indicate potential sediment deposits caused by erosion, not to be confused with tea-colored or translucent tannin-stained water that occurs naturally in coastal backwater areas and northern bogs.
Dark red, purple, blue, or black water	May indicate organic dye pollution from a leather tannery or clothing manufacturer. Also, see above.
White, milky deposits on bottoms of pools	May be a sign of aluminum pollution.
White deposits along stream banks	May indicate potential salt or brine pollution.
Foaming white suds on surface	Frequently indicates detergent pollutants from industrial, municipal, residential, or agricultural waste. Small amounts of foam are natural in a stream. Foam is of concern when it is several inches tall (like a bubble bath) and does not break apart easily.
Multi-colored, reflective, rainbow sheen on the surface	May indicate oil pollution. Some algae also produce an oily sheen on the water's surface. To tell the difference, poke the sheen with a stick. Oil pollution will reform, while algae sheen will break apart.

ODOR	POTENTIAL CAUSE
Rotten eggs	May be a sign of sewage pollution. May also occur naturally in marshes, bogs, or other wetlands along streams when waterlogged microbes release sulfur gas through digestion and respiration.
Musky odor	May indicate the presence of untreated wash water, sewage, or livestock waste. Decaying algae also generates this odor.
Bitter smell	May indicate presence of pesticides or metal pollutants.
Chlorine	May indicate over-chlorination by a sewage treatment plant or chemcal industry, or discharge of water from a swimming pool.

The SOS method involves collecting a sample of macroinvertebrates from a stream, identifying these organisms, and then rating the water quality (Figure 49). Water quality ratings of excellent, good, fair, and poor are based on the pollution tolerance levels of the organisms found and the diversity of organisms in the sample. In a healthy stream, there is a great diversity of macroinvertebrates, with no one organism making up the majority of the sample. The types of aquatic organisms found in a healthy stream include pollution-sensitive and pollution-tolerant organisms, with the majority from the pollution-sensitive range. The SOS Stream Quality Survey and instructions for macroinvertebrate monitoring can be found at *www.iwla.org/sos*.

Fish

Like macroinvertebrates, different species of fish have different tolerances to pollution. Many healthy streams support a wide variety of fish. If the stream lacks a healthy assortment of macroinvertebrates, carp or other pollution-tolerant fish may dominate. Pollution-sensitive fish, such as trout, small mouth bass, and sunfish, need a varied abundance of macroinvertebrates to thrive. These pollution-sensitive fish, like the insects they feed on, require good quality water and a habitat conducive to their survival.

There are several ways to collect information about fish populations in a stream. Fish can be trapped in nets or stunned (also called electroshocking) by using a machine that sends an electric current through the water. Fish are stunned for a few minutes, float to the surface and

Figure 49. Biological monitoring provides a fun, hands-on way to assess water quality. (Izaak Walton League of America photograph)

can be collected with nets. The shock is non-lethal and the fish usually recover after 10 to 20 minutes. Local anglers provide another source of information about fish populations. Also, most states maintain lists of species known to occur within specified areas or streams. Monitoring fish alone may not be as effective as monitoring macroinvertebrates because fish can move away from pollution problems while benthic macroinvertebrates are relatively immobile. For assistance with fish monitoring, contact the US Fish and Wildlife Service, state fish and wildlife agency, state or local environmental agencies, or local conservation districts. Inquire with state agencies to find out if the stream is stocked artificially with fish. Usually, only good quality streams are stocked, but the presence of stocked trout

Table 2. Habitat Requirements for the White Perch, *Morone americana.*[1]

Substrate	Compact silt, sand, mud, clay
Salinity (ppt)	Tolerate 0 to 8, optimum 0 to 1.5
Temperature (°C)	Tolerate 11 to 30, optimum 12-20
pH	6.5 to 8.5
Dissolved oxygen (mg/l)	>5
Turbidity (NTU)	<50
Suspended solids (mg/l)	<70

[1] Compiled from data in Chesapeake Bay Program, 1998.

in a stream does not necessarily indicate good water quality.

In addition to serving as a water quality indicator, the presence of fish in a stream also provides information about the stream's habitat (Table 2). Although all fish require dissolved oxygen and food, species such as trout, salmon, and small mouth bass also need good water quality and cool temperatures. Trout require spawning areas with gravel and cool, free-flowing, well-oxygenated water. After hatching, baby fish, known as **fry**, need suitable nurseries — slow moving, shallow areas of water adjacent to riffles, with shade from bank vegetation and an abundance of aquatic insects. The fry's small size enables it to conceal itself in shallow water and under roots and logs. As fish grow in size, they require more water, usually in the form of deep pools, to provide protection from predators. These older fish reside in straight reaches and pools, where they are close to food gathering areas and protective cover. Bigger fish, such as largemouth bass, require even deeper waters.

Many fish restoration programs have focused on stocking streams with fish to counter dwindling fish populations. However, if water quality or habitat loss is the reason for low fish numbers, stocking programs will be ineffective because the real problem — habitat degradation — is not addressed. The best way to help increase fish populations is to improve habitat and water quality. Projects to enhance fish habitat may include providing cover, stabilizing stream banks to prevent erosion, or increasing in-stream habitat. The cumulative effects of these projects could significantly enhance the natural diversity in populations of fish and macroinvertebrates.

A lack of biological life may indicate the presence of toxic chemicals in the water. Toxic effects fall into two categories: acute and chronic. Acute effects occur during short-term exposure to a chemical and result in death of some or all of the organisms. **Chronic toxicity** occurs when a pollutant causes harmful effects from either a single or long-term exposure to the pollutant. Chronic toxicity may or may not be lethal. If the concentration of toxic chemicals is low, the effects of exposure may not appear for quite some time. The most common lethal effect of chronic toxicity is the inability of the exposed organism to produce viable offspring. Typically, harmful but non-lethal effects include behavioral changes (e.g., decreased ability to swim), physiological changes (e.g., lowered resistance to disease and limited growth), or biochemical changes (e.g., altered blood enzyme levels). Some of these changes can affect an organism's ability to escape from predators or locate food, and may lead to death.

It also is important to note how birds and other wildlife are using the stream and its corridor. More information on documenting wildlife use is provided later in this chapter under "Map Uses of the Land" (pp. 53–55).

Chemical Testing

Chemical monitoring can be used as a supplement to biological monitoring. When a pollution problem is detected by a lack of aquatic organisms, a chemical analysis of the water may help to pinpoint the cause of the problem. Chemical monitoring involves taking a sample of the water and analyzing its chemistry to search for the presence of abnormalities (Figure 50). Volunteers commonly measure temperature, pH, dissolved oxygen, biological oxygen demand, **turbidity**, total dissolved solids, nitrates, phosphates, and fecal coliform **bacteria**. In addition, some state agencies recruit volunteers to collect water samples for analysis at accredited laboratories.

The disadvantage of using chemical monitoring alone is that it only provides information about the quality of the water at the moment the sample was taken. A pollution event may go unnoticed by chemical monitoring if it occurs in flowing water several days before a sample is taken. The results are affected greatly by temperature, time of day, and recent rainfall. In addition, because of the expense and difficulty involved, volunteers generally do not monitor for toxic substances such as heavy metals or organic chemicals like pesticides and herbicides. When used alone, chemical monitoring needs to take place at least once a week at the same time of day and in the same location for several months to provide a good indication of water quality.

There are several tests volunteers can use to gain insight about toxicity problems in a stream. One simple test is to determine the acidity (pH) of the stream. The values that represent pH range from 1 (most acidic) to 14 (most basic). The pH range that occurs naturally in most streams is between 6 and 9, which represents the pH

Figure 50. Chemical monitoring requires careful quality control. (Delaware Nature Society photograph)

range most fish need to survive. If the pH is below 6, drainage from mines or some other form of **acid discharge**, such as acid rain, may be affecting the stream. Values above 9 may indicate the presence of industrial wastes, although during warm weather, natural plant photosynthesis may cause pH to rise slightly above 9. The type of bedrock present in the stream may also affect pH. For example, limestone bedrock can raise pH values. Acidity levels fluctuate at various times, such as after a rainstorm or during low flow.

Determining temperature and dissolved oxygen content are other simple chemical measurements that can be used to evaluate water quality conditions. Most fish and macroinvertebrates require low temperatures to survive because as temperature rises, less dissolved oxygen is present in the water. Also, biological life is sensitive to rapid temperature changes. Streams that do not have a vegetative buffer may go through rapid increases and decreases in temperature as sunlight changes throughout the day. Thermal stress and shock can result when temperatures change more than one or two degrees within a 24-hour period. Although the amount of oxygen required varies among different species at different life stages, dissolved oxygen rates of five to six parts per million (ppm) are required for growth and activity for most organisms. Dissolved oxygen levels below three ppm are stressful for most aquatic life.

Some simple, low-cost chemical testing kits are available that measure dissolved oxygen, nitrates, phosphates, fecal coliform (may indicate sewage contamination or livestock waste material), and pH. More information about available chemical test kits is listed in the Watershed Stewardship Resources available on-line at *www.iwla.org/sos/resoures.*

Resources

The *Water Quality Standards Handbook — Second Edition,* available on the Environmental Protection Agency web site *www.epa.gov/waterscience/standards/handbook/index.html,* provides guidance on the national water quality standards program. Environmental Protection Agency regional offices and states may have additional guidance that provides more detail on selected topics of regional interest. State environmental protection or water resources agencies may have stream-by-stream information on water chemistry and flows.

The Environmental Protection Agency also manages the national on-line STORET (STOrage and RETrival) database program, which has water quality, quantity, and biological information from

---

Tip Box

Each state has a program to set standards for the protection of each body of water within its boundaries. These standards set limits on pollutants and establish water quality levels that must be maintained for each type of water body based on its designated use. The Environmental Protection Agency's Water Quality Criteria Program (*http://www.epa.gov/waterscience/standards/*) provides more information on state standards and the Fish and Wildlife Service's Habitat Suitability Index Models (*http://www.nwrc.usgs.gov/wdb/pub/hsi/hsiintro.htm*) provides habitat information for evaluating impacts on fish and wildlife habitat resulting from water or land-use changes.

800,000 sites. STORET can be accessed on-line at *www.epa.gov/storet.* The Global Rivers Environmental Education Network (GREEN) hosts a web site *(www.green.org)* to provide water monitors and students with a place to store their water monitoring data, track their water monitoring projects, and to provide educational resources to successfully implement a school-based water monitoring program. The web site is designed as a place to store water monitoring data, allow others to see what has been found, to compare the work of one group to that of others in the area, and to examine water quality trends across the country. Once a project page has been established, groups can use this site to track all of the monitoring data as long as they continue the water monitoring effort.

State fish and game agencies may have information on water quality and quantity and their biologists can often help volunteers to evaluate the data. The US Geological Survey is another good source of information about ground and surface water; the agency produces reports for each state *(http://water.usgs.gov/nawqa/)* and these might include data from stations on or near your stream. Also, look for water quality information from state and federal land managers (such as the Forest Service, Bureau of Land Management, and the Federal Highway Administration) and local universities.

Volunteers can easily walk the stream and recognize unusual colors and odors. Macroinvertebrate monitoring also is an activity routinely accomplished by volunteers. Some help with the identification of macroinvertebrates may be needed. Fish monitoring through electro-shocking requires expert assistance. Volunteers can

Figure 51. Mapping each wildlife and human land use along the stream corridor during the design of a stream restoration project ensures that all users of that part of the ecosystem are accounted for. (Izaak Walton League of America photograph)

use simple chemical monitoring test kits with ease. For complex measurements, seek technical team assistance.

Map Uses of the Land

It is important for any enhancement project to note the ways in which people and wildlife use the stream, the riparian area, and the surrounding land (Figure 51). Information on human land uses can identify new partners for the project and new opportunities to expand the enhancement work into the surrounding watershed. It is also important to keep the needs of all of the stream's users in mind when planning a stream corridor project. Uses by riparian wildlife may reveal ways to preserve and improve the benefits of the riparian corridor to a variety of wildlife.

One way to determine land uses in the watershed is to examine maps of the area. Topographic maps, watershed maps, and street maps all can help in identifying land uses surrounding a stream. Another important step in inventorying land uses is to walk the stream and its watershed and record the locations of each land use. Note areas of agriculture, development, grazing, forestry, mining, parkland, undisturbed forests, etc.

Wildlife Uses

In addition to the macroinvertebrates and fish that spend all or most of their life cycles under the water, there are many other animals that may depend upon the stream's habitat for survival. Amphibians, reptiles, birds, and mammals rely on the water, plants, and soil that make up the riparian corridor. If the project goals include improving wildlife habitat, it is particularly important to record information about the wildlife in the watershed. Quietly observe the stream corridor and note the presence of birds, mammals, reptiles, and amphibians. Look for nests, tracks, trails, scat, and other signs of wildlife use. Record this information on the site map, in a field notebook, and through photographs.

Urban Development and Related Uses

A first step in identifying development and related land uses surrounding the stream can be accomplished by examining topographic maps and requesting a copy of the current planning or **zoning** map. Local zoning maps will indicate what type of future development is allowed in what areas. Also investigate any water developments and diversions in your watershed and how they are being managed. If you live in an urban area, find out how the input and output from the local water and sewer authority affects the streams and rivers in your area.

Collect information on any utilities, water pipes, or sewers along the stream. Utility pipes for telephone cables, gas lines, electric cables, and other buried service lines should be noted so they are not disturbed during any enhancement project. Round metal lids found along the floodplain labeled "sewer" indicate the presence of sewer pipes. Contact local government planning offices for maps showing the location of other underground utilities. Also, note the locations of any pipes discharging into the stream. To determine that all point-source dischargers have current permits and that they are in compliance with their permits, contact the water regulatory agency to obtain copies of their discharge permits and compliance records.

Agricultural Uses

Note areas where livestock graze along streams or cross the stream. Livestock can damage stream banks during crossings and often increase erosion problems. If livestock routinely cross the stream site, consider enhancement options such as stream bank fencing, installing bridges or other crossings for livestock, and making provision for livestock watering. Livestock grazing along streams can also contribute to bank instability. Buffer plantings of shrubs and trees might help keep livestock from grazing too close to stream banks. If runoff from agriculture or other activities is polluting the stream, contact your local soil and water conservation district or other appropriate agency.

Chapter Five presents additional techniques for responsible rangeland management.

Recreational Uses

If people routinely fish, swim, or canoe at the stream site, it is important to maintain access for these recreational activities. Another consideration is how to make the site accessible to persons with disabilities. Educational signage about the project can relay the need for enhancement to the recreational users of the stream. Consider incorporating new recreational opportunities through the project, such as hiking trails or observation decks. Encourage recreational users to become involved in the project.

Volunteers can gather and record all of the information needed to map habitat and land uses for the stream site. It may be useful to enlist technical team members or others with experience in identifying animal tracks and scat. Technical team members or local government agencies also may be able to provide background information and materials.

Monitoring Results

As mentioned earlier in this chapter, information collected during the watershed assessment and during the site inventory can be used as baseline data for monitoring the stream corridor project. Monitoring provides a great deal of information important to the success of any enhancement project. Monitoring baseline conditions throughout the watershed can help prioritize sites for enhancement and set realistic goals. Monitoring during project installation can reveal any adjustments that may be needed before project completion. After the project is completed, monitoring results can be compared with baseline data collected during a site inventory to evaluate project success. Post-project monitoring also reveals any

maintenance needs that will ensure the project's success (see Appendix G).

Before beginning a site inventory or collecting any other baseline monitoring data, determine which parameters need to be monitored based on goals of the project. Project goals may include general goals such as stabilizing eroding banks and improving riparian habitat, or more specific goals such as re-establishing native vegetation on a section of stream bank. The next step is to set up permanent monitoring locations, develop a schedule for monitoring each parameter, and recruit people to do the monitoring. Seek technical team assistance to help with all of these steps and to determine a quality assurance plan for the project.

Information on setting project goals can be found in Appendix E.

Effective Monitoring

In order to monitor the project's success, the monitoring plan must isolate its effects so that data specifically addresses the effect of the project rather than other factors. Some ways to isolate the effects of the project include collecting information on the same parameters at the same monitoring stations before, during, and after the project installation. Ideally, monitoring should start at least one year before project installation and continue for at least five years after installation. Since isolating the effects of the project from other factors that may be affecting the stream's water quality is difficult, many groups choose to monitor upstream and downstream of the project site. Another way to monitor the project success is to choose a reference site and compare data from the reference site to the data at the project site.

Efficient monitoring uses existing data when possible and minimizes the number of measurements. Also, it is

important to include biological monitoring as one of the parameters because organisms show sensitivity to complex problems that are not as well detected by chemical or physical indicators alone. According to the Federal Interagency Stream Restoration Working Group, "in-stream organisms are the 'judge and jury' of project success."

Quality Assurance and Quality Control

Quality assurance and quality control are the methods and procedures used to ensure that the data collected are accurate and valid. Ensuring the array and variety of data is important both for gaining credibility within the community and for making sure that monitoring data is providing useful information for the stream corridor project.

The project team may want to develop a quality assurance project plan (QAPP) to ensure that quality data are collected. The quality assurance project plan is a document that outlines the procedures volunteers must take to ensure

that their data meet project requirements. For example, the QAPP outlines procedures such as the training requirements for volunteers, specifications and maintenance of the monitoring equipment, the method of documenting data, the management of data, and more. The QAPP is an invaluable planning and operating tool that outlines the project's methods of data collection, storage, and analysis. It serves not only to convince skeptical data users about the quality of the project's findings, but also to record methods, goals, and project implementation steps both for those involved in collecting the monitoring data and for others who may want to use the data for other projects. More information on developing a QAPP for volunteers is available in the Environmental Protection Agency's publication *The Volunteer Monitor's Guide to Quality Assurance Project Plans*, which is listed in the Watershed Stewardship Resources document available on-line at *www.iwla.org/sos/resources*.

Chapter Four

Choosing Stream Enhancement Methods and Techniques

*O*nce the relevant background and baseline information is collected through a site inventory, there are several methods to designing and implementing a stream enhancement project. All chosen techniques should help streams to readjust to the land-use conditions in the watershed. Some techniques are active and include applying direct physical changes to the bank; other techniques are passive, such as the management of disturbances like livestock grazing and forest production in the watershed, allowing the banks to adjust naturally. The methods used to design stream enhancement projects vary widely, as do the prescribed treatments. To choose the best solution, a match must be made between the project's objectives, existing site conditions, possible techniques, and protecting the existing fish and wildlife.

This chapter focuses on a variety of techniques for **stream bank stabilization**. Without proper expertise and data collec-

tion, stream bank stabilization can exacerbate erosion problems downstream. However, there are less complex stream enhancement projects that do not require as much technical expertise. Planting riparian forest buffers, for example, can be a fun, relatively simple community project that improves water quality and increases habitat for fish and wildlife (Figure 52).

The principles, techniques, and recommendations in this chapter draw information from case studies in many vastly different regions of the country. However, the methods presented in this chapter are only suggestions and will not always apply to the unique characteristics of every degraded stream, nor to the objectives and constraints of each community. Again, it is crucial to consult the project's technical team and to involve community stakeholders when designing enhancement projects.

Figure 52. Volunteer assistance can be an effective way to combat financial constraints, especially when planting riparian buffers. (Izaak Walton League of America photograph)

Types of Stream Bank Stabilization

Stream enhancement projects fall into two broad categories: those that correct the problem, and those that compensate for it. Although the most effective way to stabilize a stream bank is to eliminate the cause of the instability, measures to compensate for a problem are often used in addition to, or instead of, correcting the fundamental cause. Correcting the problem often involves sound planning for land uses in the watershed. Information about land-use planning and ways residents can encourage good community planning can be found in Appendix H.

Bank stabilization methods can be categorized into three fundamental types: **Structural**, **vegetative**, and **bioengineering**. Structural stabilization methods are those that rely on riprap and/or large boulders to anchor the **toe** (the bottom of the bank), redirect erosive flows away from a portion of the bank, or armor the entire bank with a protective shield. These methods should only be considered when vegetative or bioengineering approaches are not possible due to extreme erosion

and stream bank instability. Vegetative stabilization methods are those that use plants or plant cuttings to stabilize the bank. Bioengineering blends structural and vegetative stabilization methods by incorporating various materials (such as rock, timber, soil, and plants) to secure the bank. In combination with these materials, bioengineering methods may also include erosion control fabrics such as **jute mesh** and **coir.**

Bioengineering uses native plants and natural materials to repair unstable areas through understanding the engineering properties and capacity of plants. The goal of bioengineering is to establish diverse native plant communities capable of self-repair as they adapt and function along the stream. Bioengineering is an inexpensive and attractive way to control erosion, sediment, and flooding while providing educational opportunities and promoting stream stewardship. Bioengineering techniques are not recommended in all situations, especially urban areas with high stormwater flows.

When Bank Stabilization Is Not Appropriate

If a watershed is undergoing rapid changes due to urbanization, such as increased paving, conversion of forested land to agricultural land, installation of an upstream dam, or increased water withdrawals, the hydrology of a stream may be unstable. As stated earlier, streams seek to achieve a state of equilibrium. If upstream land uses have changed the amount of water flowing into the stream, the stream may be in the process of establishing a new slope or gradient. The calculations about stream flow may not be valid months down the road and enhancement projects could fail. Thus, a stream enhancement project would not be appropriate in this situation.

Frequent widening of the stream is a sign that the stream is in the process of achieving a new gradient. In this case, it is best to delay or cancel a project until a new, more stable gradient is achieved.

If information about a stream is not available from government sources, a technical team cannot be formed, or consultants are too expensive, it is best to postpone a project rather than overlook critical information and potentially harm the stream. If more information is needed, take time to gather assistance and support for a project.

If the watershed is highly urbanized and the stream receives a high volume of water during rainstorms, flow levels might be too high for certain bioengineering methods. When this high volume of water reaches streams through storm sewers, it gouges stream bottoms, makes banks unstable, and carries material downstream with great force. A long-term approach to solving this problem may be to work with city and county planners on stormwater management strategies such as diverting stormwater from large paved areas to stormwater management ponds or constructed wetlands to capture, hold, and filter water runoff. Residents may also encourage local government planners to research and develop low-impact development techniques for dealing with stormwater. Excessive urban runoff also may contribute toxic chemicals that degrade water quality to the point where aquatic life cannot survive.

Before Getting Started

The rest of this chapter provides examples of potential stream corridor solutions. Before undertaking any activities, technical expertise and guidance will be required. As previously mentioned, it is recommended that an interdisciplinary team of experts knowledgeable about local conditions be recruited as your technical team. The nature of the project will likely dictate the most suitable qualifications or experience required of the team. At a minimum, the team consisting of an engineer with experience in stream systems, an ecologist knowledgeable in fisheries and riparian ecology, and a soil scientist will generate the most successful projects. Some projects may require the specialized skills of a geomorphologist, botanist, or landscape architect. Only experts can answer technical questions, such as if a stream may need to be widened and/or terraces need to be added. For more information on establishing a technical team, please see Appendix B.

To design a successful project, it is necessary to assess the causes of instability, design solutions, predict the stream's response to the prescribed solutions, implement the project, and monitor the results. Before beginning an enhancement project, observe stream functions for at least six months, including a rainy season, and perform the watershed assessment and site inventory described in chapters Two and Three.

It is important to match the appropriate enhancement techniques with the identified problem, basing decisions on the overall findings of the site analysis. Altering the bank without regard to the stream's natural state can have many unintended effects.

Appendix I — "Obtaining a Permit" contains background information and instruction on how to obtain permits.

Also, remember to check with local authorities to make sure you are following all the requirements for your project. These activities never should be attempted without requisite permits and consultation with your local flood-control management agency.

As mentioned in earlier chapters, cost is also a major consideration when planning stream enhancement projects. The more complex the project, the more technical assistance, planning, materials, and ongoing maintenance will be required. If

Ideas on how to secure project funding can be found in Appendix J — "Funding Watershed Conservation Projects."

all of the funds are not secured, it is better not to start the project. Some items to consider when developing a budget for your enhancement project include:

• The cost of contracted labor;
• The need for, and cost of, professional consultation;
• The price of materials and tool rental;
• The cost of machinery rental for grading or other structural activities;
• Permitting fees (state, county, city);
• Long-term monitoring of the site (>2 years); and
• Maintenance/irrigation for the establishment period.

Riparian Buffers

Vegetation growing by the river in the riparian buffer provides shade and food necessary for the aquatic habitat. Establishing or expanding stream corridor vegetation is a simple and effective technique to improve water quality (Figure 53). A buffer is especially important for streams adjacent to farmlands; vegetation in these areas helps trap and break down pesticides, fertilizers, and other pollutants before they enter the stream.

The minimum size of a stream buffer depends on the region of the country and watershed features such as soil type, topography, vegetation type, precipitation, the size of the stream, and the land-use intensity. Areas with soils that are easily eroded, steep slopes, vegetation with shallow roots, and heavy land-use activity typically require wider buffer zones.

Even when these factors have been taken into account, there still is much debate about how wide the vegetated area should be along streams. A buffer that is too narrow (10 feet or less on either side) may not

Figure 53. Planting a riparian buffer is one of the most simple and effective methods of stream enhancement. (Izaak Walton League of America photograph)

More local information about buffer width requirements for streams is often found through the permitting process. In addition, state or local departments of forestry may have guidelines to protect the function of riparian buffer zones (called stream-side management areas). Also, check with the local government to see if it has ordinances that apply to landscaping, tree preservation, buffers, erosion, sediment, and stormwater management.

While the widest possible forested buffer on both sides of the stream is an ultimate goal, getting a buffer of any size is the most important first step. Buffers can be widened over time.

Vegetative Techniques

Stream bank erosion can sometimes be averted by simple measures such as preventing damage to existing vegetation, allowing damaged plant communities to recover naturally, and re-establishing

adequately filter runoff and cannot serve as sufficient habitat for riparian wildlife. The Environmental Protection Agency recommends that stream buffers have a minimum width of 35–50 feet on each side, though they might be narrower or wider depending on site-specific factors (Figure 54). The Forest Service recommends at least 100 feet for steep mountain streams. Another suggested buffer width is two to five times the width of the stream channel.

Figure 54. Vegetation located in the riparian zone protects stream health by acting as a buffer between developed land and the stream channel. (Izaak Walton League of America figure)

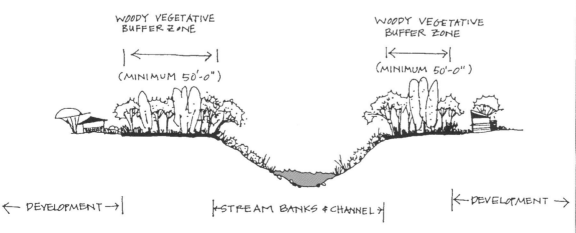

vegetation where it has been removed. Stream bank vegetation helps stabilize banks by absorbing excess water and catching sediment. Also, plant and tree roots physically hold banks in place. While vegetation can be a cost-effective, self-maintaining mechanism for improving bank stability, the plant species selected must meet the specific conditions of the site. Whether or not a plant species is appropriate at a particular site depends on several factors: purpose of planting, soil moisture (permeability and drainage), available sunlight, brush competition, potential for animal damage, and elevation. The answers to the questions that follow will provide the basis for selecting the most appropriate vegetation for a stream bank stabilization project.

- What are the specific goals and objectives of the project? Re-establishment or enhancement of existing plant community? Restoration of previous plant community?

- What are the physiographic characteristics of the project site (elevation, slope aspect, and topography)?

- What are the climatic characteristics of the project site (types, amount, and timing of precipitation; length of growing season; average temperature; velocity and direction of prevailing winds; available light)? Will the light requirements of the plant assemblage (full light, partial shade, or shade tolerant) be met under the existing and/or anticipated site conditions?

- What soil types exist in the project site and adjacent areas? What are the specific characteristics of these soils (permeability; drainage; available water capacity; fertility; texture)?

- What is the hydrology of the project site? Is the site periodically covered with water? If so, how frequently and for what length of time? Is the site covered by standing water or flowing water? If flowing water, what is the estimated depth and velocity?

- What is the condition of the existing plant community? Is it a natural plant assemblage? Has it been recently altered or disturbed? If altered or disturbed, to what extent? What was the cause of the alteration (a single event versus ongoing disturbance)?

- Are there existing or planned access roads or pathways in and near the project site? What form of vegetation (herbs, shrubs, trees) is appropriate for the intended function of the bank (such as recreation or maintenance access, if any)?

- Do site conditions require special design considerations such as vegetation height or shape, type of root structure for erosion control and bank stability, soil type and depth?

Selecting Appropriate Vegetation

    The slope, soil suitability, moisture, and sunlight availability of the site are key considerations for a successful planting project. Soils should have adequate nutrients and moisture to support plant growth. Plants probably suffer more from moisture-related problems (either too much or too little moisture) than from any other causes (Table 3). Therefore, it is important to understand the soil moisture conditions of a site by reviewing soil surveys as described in Chapter Three. A few woody plants are adapted to frequent or prolonged flooding or to poorly drained soils. Most woody vegetation, however, will grow best when planted in easily drained soil and usually does not tolerate continuous waterlogged soil conditions (Figure 55). Sites with poorly drained soils may require special treatment, such as adding soil amendments to improve water flow around the roots.

Plants, shrubs, and trees that are native to the area should be considered first for any vegetative planting project. In addition, locally obtained plants are generally better adapted than plants obtained from distant sources. Consult with local experts to determine which native plants have deep, branching roots, provide good shade and are long lived. It is also important to determine which plants root easily and can develop roots from all plant parts, including buried twigs. Examples of plants with good rooting ability are the black willow (*Salix nigra*) and the streamco willow (*Salix purpurea*). Both species also provide good wildlife habitat. The Natural Resources Conservation Service also provides a web resource called the PLANTS Database. This database (*http://plants.usda.gov/*) includes names, checklists, automated tools, identification information, species abstracts, distribu-

Additional information about how to select appropriate plants and contact information for native plant nurseries is located in the Watershed Stewardship Resources document available on-line at *www.iwla.org/sos/resources.*

tional data, crop information, plant symbols, plant growth data, plant materials information, links, references, and other information.

Regional plant inventory lists can provide valuable information on native plants best suited to the project. However, there are no substitutions for on-site analyses and site-specific recommendations. Only plants from sites with ecological conditions similar to the project site should be used. The discussion on reference reach in Chapter Two can provide clues to suitable vegetation.

The plants native to an area depend on the climate, stream hydrology, geology, alkalinity of the soil, and many other factors. In the Pacific Northwest, vegetation adjacent to streams tends to include willows (*Salix* spp.) and alder (*Alnus* spp.). Both groups will flourish when their root systems are damp. Alder has the special ability to take nitrogen gas out of the air and make it into biologically useful nitrogen (nitrogen fixation). This makes the soil near streams and on floodplains more fertile. Cedar trees (*Thuja* spp.) also like the acidity of Pacific Northwest streams and watersheds.

Willows, often in dense thickets, can also be found next to rivers in the southwestern United States. Cottonwood trees (*Populus deltoides*) grow a little farther back from the river, forming the outer rim of the corridor. Like most poplars, they like a moderate amount of water. Unlike rivers of the Pacific Northwest, those of the Southwest tend to be very alkaline. Most coniferous trees like piñon pine (*Pinus edulis*) grow outside the floodplain to keep their roots dry. There are other

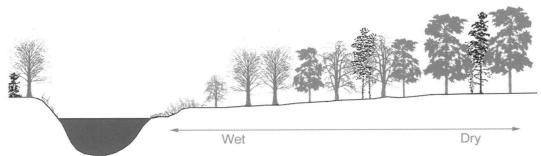

Wet   Dry

Figure 55. Vegetation that can tolerate frequent flooding and waterlogged soil can be found close to the stream edge. Less tolerant species tend to grow farther back in the floodplain. (Izaak Walton League of America figure)

Table 3. Sample Planting Recommendations According to Moisture Conditions.

WET ◄─────────────────────────────────► DRY

TREES

Silver maple	← Red maple →	White pine
Box elder	Bitternut hickory →	Black cherry
Persimmon →	← Redbud →	Sassafras
Black ash	Hackberry →	Canada hemlock
Red ash	American beech →	White oak
Pawpaw →	← White ash →	← Red oak
Sweet-bay magnolia	← Honey locust →	Chestnut oak
Sycamore →	Kentucky coffee tree →	Shagbark hickory
Cottonwood	← Sweet-gum	Sugar maple
Swamp white oak	Tulip tree →	Black walnut
Willow oak →	Black-gum	Black locust
Sandbar and black willow	← Pin oak →	Yellow poplar
Bald cypress	← Large-toothed aspen →	Eastern red cedar
	← Swamp white cedar →	
	← Loblolly pine →	

SMALL TREES/SHRUBS

River birch	← Black/sweet birch →	Hop-hornbeam
Smooth alder	Mountain laurel →	Witchhazel
Red chokeberry	← Hornbeam →	Staghorn sumac
Black chokeberry →	Yellow birch	Nannyberry
Groundselbush	← Shadbush →	Blackhaw
Red osier and silky dogwood	← Gray and flowering dogwood →	
Summersweet →	Fringe tree →	
Winterberry →	American hazelnut →	
Inkberry →	← Black huckleberry	
Swamp rose	← Common spicebush →	
Swamp azalea	← Rosebay rhododendron	
Meadowsweet →	← Southern arrowwood	
Highbush blueberry →	← Ninebarr	
Witherod →	← American elder	
Northern arrowwood	Bayberry →	
	← Highbush cranberry →	
	Red elm	
	← American holly →	
	← Buttonbush →	

Note: Arrows denote that certain species can tolerate either a wetter or drier environment.

conifers, however, like alligator juniper (*Juniperus deppeana*) that will grow quite close to water.

For the past several decades, grass filter strips have been recommended as streamside vegetation. However, grass does not make the best buffer because its roots are shallow — only five to six inches deep — and therefore do not deeply anchor the soil along the bank. Grass also does not provide the habitat values that trees and shrubs do (Table 4). However, if grass is the native vegetation for your stream's banks, such as is the case in the tallgrass prairies of the Midwest, then you might want to encourage some of this type of vegetation along your stream. Tall grasses such as river bulrush (*Scirpus fluvialtilis*), prairie cord grass (*Spartina pectinata*), and Canada bluejoint (*Calamagrostis canadensis*) tend to hang over the stream and provide shade and shelter. Projects can combine plantings of these grasses with elderberry (*Sambucus canadensis*), dogwood (*Cornus* spp.), and shrubby willows, such as peachleaf (*Salix amygdaloides*) and sandbar willows (*Salix exigua*).

Cuttings, seedlings, container plants, and bare-root plants can be used to stabilize banks and expand riparian buffers. Vegetative cuttings are live plant materials (twigs and branches) that can be placed in the ground to root and grow. They work well in difficult planting situations such as rock slopes and boulder outcroppings. They work best when there is a local source of easily rooting deciduous plant material. Cuttings should be 1/2 to two inches in diameter and range in length from two to six feet. Tools such as chainsaws, bush axes, loppers, and pruners may be needed to collect cuttings. Transport plants under a tarpaulin to avoid drying or budding, and store them in a cool, dark place until ready for use. Vegetation should be planted within eight hours of harvesting. Be sure to obtain permission before doing any cutting and only collect the amount of plant material that is needed. Also, don't cut more than 1/3 of the host plant to ensure that the host plant survives. For more information on live cuttings, see Appendix K.

Choosing the Right Time to Plant

The ideal time to plant cuttings is during the dormant season. It also is possible to plant cuttings during the start of the growing season in early spring or late fall before the first hard freeze. In this case, it is ideal to store them in a large commercial cooler to keep them from budding too soon. Consult with a plant materials specialist to determine the proper temperature for storage. The cuttings can be removed on the day of installation. If a commercial cooler is not available, be sure to store the cuttings in a cool, dark place in moist soil or clean water. Do not plant cuttings during the heat of the summer months.

Unfortunately, timing does not always coincide with in-stream permitting requirements or construction projects that may be occurring in the area. Always follow the timing requirements of in-stream permits if your project needs a permit. If construction upstream of the site will occur in the summer, postpone the planting of vegetation until the fall. Seedlings should be planted in the spring, keeping in mind that nursery-grown stock is heartier than wild seedlings. To help ensure plants will thrive, use native plants derived from local stock. If planting grasses, remember that they should not be planted at the same time as woody shrubs

Table 4. Wildlife Use of Selected Species.[1]

COMMON NAME	BOTANICAL NAME	VALUE
Maple	*Acer* spp.	Moderate
Alder	*Alnus* spp.	Moderate
Bearberry	*Arctostaphylos* spp.	Moderate
Serviceberry	*Amelanchier alnifolia*	Moderate
Oregon grape	*Berberis nervosa*	Moderate
Paper birch	*Betula papyrifera*	Moderate
Red-osier dogwood	*Cornus stolonifera*	High
Hazelnut	*Corylus cornuta*	High
Salal	*Gaultheria shallon*	Moderate
Oceanspray	*Holodiscus discolor*	a
Trumpet honeysuckle	*Lonicera ciliosa*	Moderate
Black twinberry	*Lonicera involucrate*	Moderate
Crabapple	*Malusfusca* spp.	Moderate
Indian plum	*Oemleria cerasiformis*	Moderate
Mock orange	*Philadelphus lewisii*	a
Pacific ninebark	*Physocarpus capitatus*	a
Sitka spruce	*Picea sitchensis*	Moderate
Lodgepole pine	*Pinus contorta*	High
Western white pine	*Pinus monticola*	High
Black cottonwood	*Populus balsamifera*	High
Quaking aspen	*Populus tremuloides*	Low
Bitter cherry	*Prunus emarginata*	High
Chokecherry	*Prunus virginiana*	High
Douglas fir	*Pseudotsuga menziesii*	Moderate
Ferns	*Pterophyta* spp.	Low
Cascara	*Rhamnus purshiana*	Moderate
Currant	*Ribes* spp.	Moderate
Rose	*Rosa* spp.	Moderate
Salmonberry	*Rubus spectabilis*	High
Blackberry	*Rubus* spp.	High
Thimbleberry	*Rubus parviflorus*	High
Willow	*Salix* spp.	High
Elderberry	*Sambucus* spp.	High
Sitka mountain ash	*Sorbus sitchensis*	High
Hardhack	*Spiraea* spp.	Moderate
Snowberry	*Symphoricarpos albus*	Moderate
Creeping snowberry	*Symphoricarpos mollis*	Moderate
Western red cedar	*Thuja plicata*	Moderate
Western hemlock	*Tsuga heterophylla*	Moderate
Mountain hemlock	*Tsuga mertensiana*	Moderate
Huckleberry	*Vaccinium* spp.	Moderate
Highbush cranberry	*Viburnum opulus*	Moderate

[1] Information in this table is from Hanley and Kuhn, 2003.
a Not all species were rated for value, only noted that they were of value. Values include nesting, resting, and feeding for birds, mammals, and other animals.

or seedlings because they tend to spread quickly, choking out the woody plants.

Container plantings are nursery stock plants that are available in a variety of sizes. Container plants can be purchased when in bloom or bearing fruit to help project managers see exactly what the completed restoration site will look like. See Appendix L for more information on using container plants in stream enhancement projects.

Bare-root plants are grown in the ground, then dug out and transplanted to another site. Bare-root plants are typically half as expensive, and usually larger and healthier than container plants, whose root systems have been confined. Bare-root plants are only available in the dormant season and must be planted soon after being taken out of the ground. In winter and early spring, bare-root plants are available at many retail nurseries or from mail-order nurseries. When buying plants, make sure the roots are fresh and plump. See Appendix M for more information on using bare-root planting in stream enhancement projects.

Plant Maintenance

The spring and fall are also active times for beaver and deer. These mammals love fresh newly planted trees and cuttings, so be sure to temporarily fence vegetation until it is well established. Another option to consider is biodegradable tree shelters, which are easily installed and removed and are relatively inexpensive for small areas (see Appendix N).

High water velocities or severe drought conditions might limit the success of installing vegetation. Temporary irrigation may need to be added to the enhancement plan if conditions are too adverse for young plants. Take the time to design a system that works, which might mean making time to water plants by hand. In arid climates, use drought tolerant species such as box elder (*Acer negundo*) or arrowleaf cottonwood (*Populus angustifolia*). There should also be adequate light at the site, especially if considering species such as willow and dogwood. Consider planting shade-tolerant plants in areas of lower light.

Removing Non-Native Vegetation

The control or removal of non-native vegetation may be an important part of any stream enhancement strategy. Non-native vegetation often prevents growth of native plants that are needed by riparian wildlife. For example, multiflora rose is common along stream corridors in the eastern United States, but it is imported from Asia. Multiflora rose is a problem for streams because it has a shallow root structure, which does little to stabilize stream banks. It also has dense thorns, blocks stream access, and provides little food value for wildlife.

Consider Bank Slope

If a vertical bank is eroded, slumped, or undercut by the water current, then planting vegetation alone probably will not work because the bank will collapse before the roots become firmly established. Enhancement projects for these types of sites might require grading the bank to a more gradual slope. For example, the gradient (steepness) of undercuts or **slumping** banks must be reduced to a slope of two feet of horizontal distance to one foot of vertical rise, a 0.5, or 2:1, grade. A slope of 0.33, or 3:1, is even better, although it might not be practical where space is limited. In some cases, steep banks can be a natural and stable

condition, especially along streams with certain bank materials. Consult with the project's technical team to determine the need for grading.

Slope grading projects that do not require moving an excess of soil can be accomplished using hand tools, such as shovels and hoes. Projects of larger magnitude, however, require the help of backhoes. Both of these projects need the assistance of the project's technical team. A large project can be expensive unless the labor is donated by a local contractor or government agency. Bank shaping should be done as close to the actual planting date as possible to prevent opportunities for increased erosion. Enhancement methods that require extensive bank sloping may be limited by the close proximity of structures (e.g., buildings, roads, utilities), loss of vegetation (e.g., large trees), land acquisition, or easements. In these situations, bioengineering techniques designed by professionals could be used to protect a steeper slope.

Bioengineering Techniques

The following pages include descriptions of various bioengineering techniques. These descriptions are not to be construed as step-by-step instructions for installation. As stated throughout this handbook, it is essential to include the project's technical team in all aspects of the project. The following descriptions are intended to help citizens become informed contributors in planning stream enhancement projects.

In addition to planting grasses, shrubs, and rooted plant stock, bioengineering techniques, such as **live stakes**, **fascines**, and brush layers, can be

Please see appendices O and P for more information on watering and weeding for plant maintenance.

used to stabilize stream banks. Structural protection methods such as riprap, **deflectors**, and **gabions** can also incorporate vegetation to increase the structural integrity of a bank. Because many of these techniques mimic natural processes, they are more likely to be self sustaining once established. Although their installation is labor intensive, they are often useful on sites where machinery use is limited or not feasible.

Live Stakes

Live stakes are woody cuttings that can root when planted and are large enough to provide support when vertically tamped into the ground. Also referred to as pole cuttings, they are generally cut from trees that are two or more years old. Cuttings from plant species that root easily will grow if planted under favorable conditions. Most willows, red osier dogwood, many poplars, and cottonwoods easily grow from cuttings set in moist soil (Figure 56).

Live stakes can be used alone or with straw, jute mesh, and coir (a coconut-fiber mesh) to provide surface protection or repair small slumps and gullies. Larger live stakes can be very effective in stabilizing eroding toes or slopes of banks (Figure 57). They can also be installed between other bioengineering techniques. Live stakes are effective when:

• Construction time is limited;
• An inexpensive method is necessary;
• The problem is very simple; or
• Work in the stream channel is not allowed or desirable.

Use plants from two to 10 feet long with diameters ranging from 1/2 to four inches, depending on the species of plant. Side branches should be removed and bark

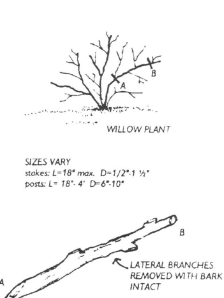

WILLOW PLANT

SIZES VARY
stakes: L= 18" max. D=1/2"-1 ½"
posts: L= 18"- 4' D=6"-10"

A   B

LATERAL BRANCHES
REMOVED WITH BARK
INTACT

BASAL (BUTT) END CUT AT ANGLE
PLANT THIS END IN SOIL

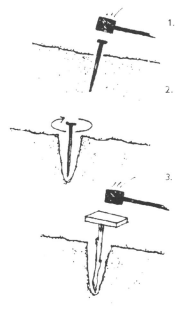

1. USING A SUBSTANTIAL SLEDGE HAMMER AND CONSTRUCTION STAKE (METAL POLE W/POINT) MAKE PLANTING HOLE

2. WIGGLE STAKE LOOSE AFTER EVERY FEW BLOWS OF HAMMER SO STAKE CAN BE REMOVED AFTER MAKING HOLE AS DEEP AS POSSIBLE

3. PLACE LIVE POLE IN HOLE STARTED BY STAKE. PLACE BOARD ON TOP OF POLE AND HAMMER IN 2/3 OF POLE IN GROUND (BOARD PROTECTS FROM SPLITTING. IF SPLIT OCCURS DISCARD POLE). FILL HOLE WITH SOIL AND WITH FOOT COMPACT SOIL AROUND STAKE

Figure 56. Live stakes, often readily available and at low cost, are easily prepared and placed into the ground. (Smith, O'Connor, and Uchiyama, 1995)

should be left intact. Cut the basal ends, or the base of the branch, at an angle for easy insertion. Insert three-quarters of the length of the cutting into the soil at an angle of 10 degrees or more. The growing end should be cut blunt. Install live stakes two to three feet apart using triangular spacing, with stakes pointing downstream. Firmly pack the soil around the stake.

Branch Packing

Branch packing involves layering dead branches and mulch in the gully of a small eroding section of a bank and fastening them down with live stakes and wire (Figure 58). Also called brush mattresses, they can be used both above and below the water level to help repair gullies and washed out stream banks. Large

numbers of branches are required, but cuttings often are available along another stream section or nearby tributary. Branch packing produces an immediate, natural barrier that redirects water away from washed-out areas, preventing erosion problems from growing larger. These

Figure 57. Live cuttings can be covered with damp burlap to prevent drying. Drying is a major threat to the survival of live cuttings during transport and storage. (Federal Interagency Stream Restoration Working Group, 1998)

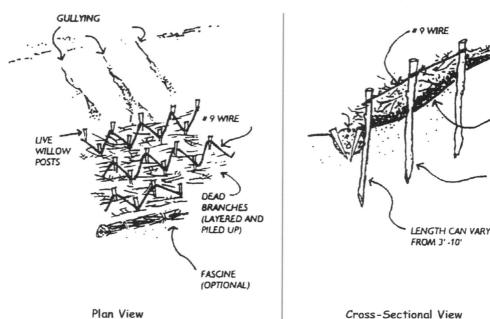

GULLYING

LIVE WILLOW POSTS

9 WIRE

DEAD BRANCHES (LAYERED AND PILED UP)

FASCINE (OPTIONAL)

Plan View

9 WIRE

DEAD BRANCHES

MULCH AND LEAF LITTER

LIVE WILLOW POSTS

LENGTH CAN VARY FROM 3'-10'

Cross-Sectional View

Figure 58. Branch packing is a way to reduce erosion problems along washed-out streambanks. (Smith, O'Connor, and Uchiyama, 1995)

structures are most suitable in low gradient streambeds where open pools are present.

Brush Layers

Brush layers are alternating layers of live branches and soil on successive horizontal rows or contours in the stream bank. They are used to stabilize a slope that is eroding by providing an immediate protective surface. The bank is cut back and a series of horizontal terraces is created. Live and dead plant cuttings, varying in length from one foot to more than six feet, are placed on each terrace. The tips of the live cuttings face toward the stream and the basal ends are shoved into the slope (Figure 59). The buried portion of the branches takes root to form a permanent reinforced installation, while the tips produce vegetative top growth. Brush layers are used to provide slope stability and trap soil, decaying leaves, and other debris. They also help to dry excessively wet slopes and aid infiltration of water on dry slopes. They are best used

when the slope of the bank is steep and eroding (Figure 60). Erosion control fabric, such as coir or jute, can also be used to enhance the stabilization.

Live Fascines

Live fascines (also called **wattlings**) are sausage-like bundles of live woody cuttings tied together and inserted into a shallow trench dug into the stream bank. They are held in place with live or dead stakes. Often used with other vegetation, fascines protect banks from erosion, particularly at the stream's edge where water levels fluctuate. Because materials usually are available locally, fascines are an economical method to control erosion. In addition, this method offers immediate protection that grows into a durable, natural stream bank stabilizer. Fascines need to be protected from cattle or other grazing animals so that they are not eaten or trampled.

Fascine bundle lengths will vary depending on the size of the eroding surface and the cuttings available, but

Installation of Brush Layers

PLANT CUTTINGS LAID ON TOP OF EXCAVATED TERRACE. BUTT ENDS ARE PLACED INTO BANK

TYPICAL HORIZONTAL SPACING OF BRUSH LAYERS

bottom of slope: 3-5' between layers
middle of slope: 5-8' between layers
top of slope: 8-12' between layers

A

TERRACE

EDGES OF FILL

TERRACE

BUTT ENDS

A'

BEFORE TOPSOIL COVERING

AFTER TOPSOIL COVERING

PLAN VIEW

A ——— FILL ——— A'

BRUSH LAYER

TERRACE

SLOPE BACK TERRACE

CROSS-SECTIONAL VIEW A-A'

Brush Layering with Wooden Stakes or Live Pole Cuttings

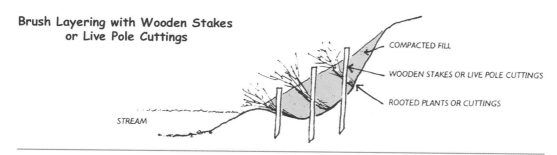

COMPACTED FILL

WOODEN STAKES OR LIVE POLE CUTTINGS

ROOTED PLANTS OR CUTTINGS

STREAM

Brush Layering in Combination with Other Techniques

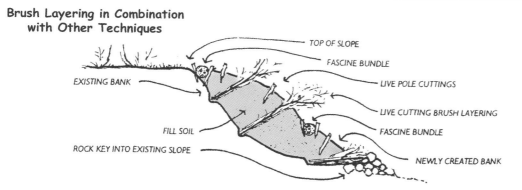

TOP OF SLOPE

FASCINE BUNDLE

LIVE POLE CUTTINGS

LIVE CUTTING BRUSH LAYERING

FASCINE BUNDLE

NEWLY CREATED BANK

EXISTING BANK

FILL SOIL

ROCK KEY INTO EXISTING SLOPE

Figure 59. Brush layers are an effective stabilization technique for steep slopes. Using live poles further reinforces the brush layers. Cuttings will take root and anchor the soil in place. Other techniques, like live pole cutting and fascine bundling, can be used in combination with brush layering. (Smith, O'Connor, and Uchiyama, 1995)

typically are about four feet long. Cuttings must be from species that root easily and have long, straight branches, such as young native willows, cottonwoods, or red osier dogwood. The cuttings are placed in bundles about eight inches wide and tied together with twine every 12 to 18 inches with the basal ends in alternating directions (Figure 61A). Shrub willows are highly desirable for bank stabilization

Figure 60. This volunteer is using the brush layering technique to stabilize a degraded stream bank. (Izaak Walton League of America photograph)

because they branch outward. The branches contained in the bundles should be at least four feet long, with a maximum diameter of one inch.

Before using fascines, the toe of the bank may first need to be stabilized. To do this, drive wooden stakes in at the base of the bank, beginning at average low-water level and moving in a line along the contour of the slope. Rocks can also be shoved in below a fascine to prevent undercutting. Next, dig a shallow trench as deep as the diameter of the fascines, no more than one hour before planting. Water the trench just before and after installation to prevent the trench from drying out. Place the fascines in the trench with each bundle overlapping the next by a few inches. Push the ends of the bundles into each other to lock together and create a continuous strong structure (Figure 61B). Drive stakes, spaced one foot apart, through the bundles to secure them, with extra stakes at the joints. Cover the fascines with the soil previously removed from the trench (Figure 61C). Stakes should protrude from the soil at least six inches. Finally, tamp down the soil. This

sequence proceeds row by row up the contour of the slope, with trenches about three feet apart (Figure 62).

Installation should occur when water levels are low, beginning at the base of the bank and continuing up the slope. Fascines work well in areas where heavy equipment is restricted, where existing vegetation could be damaged with machinery activity, and on rocky sites where digging is difficult. Fascines are a good solution when project goals include habitat enhancement and an aesthetically natural look, and where the budget is limited.

Live Crib Wall

A live crib wall is a rectangular framework of logs, rocks, and live cuttings designed to protect an eroding stream bank or to prevent formation of a split channel. It is best used on the outside bends of main channels, where currents are strongest (Figure 63). Live crib walls can be complicated to install and expensive if materials are not available locally; however, they provide long-term durability with a natural look as live cuttings grow. This method should not be used where the streambed is eroding, because the water will continue to erode the bed below the crib wall level, exposing the bottom of the crib wall to erosion.

Dig the crib wall base two to three feet below the existing streambed. Excavate the stream bank and place a log parallel to the excavated bank. Place the next series of logs at right angles to the

Figure 61. When properly designed and installed, live fascines can provide immediate, long-term protection from stream bank erosion. (A) First, bundles of live woody cuttings are tied together. (B) Then, bundles are pushed together to create a continuous, strong structure. (C) The linked fascine bundles are then placed in a shallow trench, held in place with stakes, and covered with soil. (Izaak Walton League of America photographs)

first log. The ends of each log overlap the perpendicular log below. Secure each log by cutting notches in the wood. Drill holes through the overlapping logs and use steel pins to hold them securely. Fill the openings with cuttings and soil. The top layer is composed of compacted soil, and the lower end of the crib wall is protected by riprap.

Cribbing should be embedded at least two to three feet below the streambed and extend to at least half the height of the bank, using bark-free logs. They should be tied well into the bank and planted with vegetation at upstream and downstream ends. As with most enhancement methods, crib walls work best when used with vegetation.

Tree Revetments

Tree **revetments** are medium- to large-cut evergreen trees which are anchored into an eroded stream bank to slow erosion and provide a place for sediments to deposit. One or more trees are secured in a row along the toe of the bank. When a one-row revetment is not adequate, additional rows of trees may be installed on top of the first row to protect the bank. Trees are placed with the trunk end upstream and are secured to the bank with cables. Wrap the cable around the trunk above the bottom limbs or thread it through a hole drilled in the trunk and secure with a clamp. Anchor the other end of the cable into the bank near the toe with a disk anchor, T-post, or duckbill earth anchor (all are available at hardware stores or farm supply stores). Use another cable and anchor to secure the top part of the tree. Pull the cables tight to hold the tree close to the bank. Work upstream and place additional trees so that each tree overlaps the previously placed tree until the revetment extends beyond the eroded area.

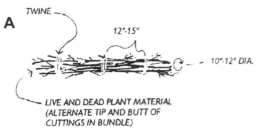

A TWINE

12"-15"

10"-12" DIA.

LIVE AND DEAD PLANT MATERIAL
(ALTERNATE TIP AND BUTT OF
CUTTINGS IN BUNDLE)

PREPARE WATTLING BUNDLES OF LIVE AND DEAD PLANT
MATERIAL WITH BUTTS AND TIPS ALTERNATING. WATTLES
ARE 10"-12" IN DIAMETER, TIED 12"-15" ON CENTER. USE
PLANTS THAT ROOT EASILY FOR LIVE CUTTINGS.

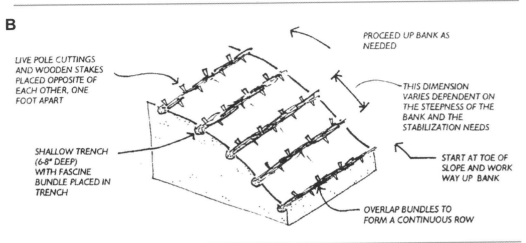

B

LIVE POLE CUTTINGS
AND WOODEN STAKES
PLACED OPPOSITE OF
EACH OTHER, ONE
FOOT APART

SHALLOW TRENCH
(6-8" DEEP)
WITH FASCINE
BUNDLE PLACED IN
TRENCH

PROCEED UP BANK AS
NEEDED

THIS DIMENSION
VARIES DEPENDENT ON
THE STEEPNESS OF THE
BANK AND THE
STABILIZATION NEEDS

START AT TOE OF
SLOPE AND WORK
WAY UP BANK

OVERLAP BUNDLES TO
FORM A CONTINUOUS ROW

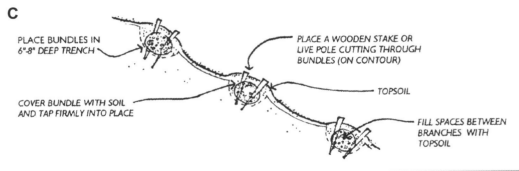

C

PLACE BUNDLES IN
6"-8" DEEP TRENCH

COVER BUNDLE WITH SOIL
AND TAP FIRMLY INTO PLACE

PLACE A WOODEN STAKE OR
LIVE POLE CUTTING THROUGH
BUNDLES (ON CONTOUR)

TOPSOIL

FILL SPACES BETWEEN
BRANCHES WITH
TOPSOIL

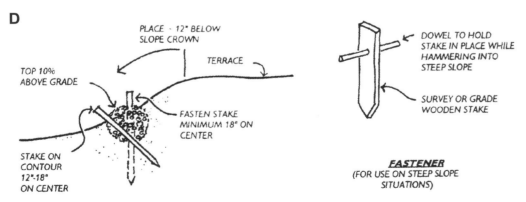

D

PLACE - 12" BELOW
SLOPE CROWN

TERRACE

TOP 10%
ABOVE GRADE

FASTEN STAKE
MINIMUM 18" ON
CENTER

STAKE ON
CONTOUR
12"-18"
ON CENTER

DOWEL TO HOLD
STAKE IN PLACE WHILE
HAMMERING INTO
STEEP SLOPE

SURVEY OR GRADE
WOODEN STAKE

FASTENER
(FOR USE ON STEEP SLOPE
SITUATIONS)

Figure 62. (A) Branches for fascine bundles should be tied so that their growing tips face opposite directions. (B) Install fascine bundles in shallow trenches parallel to the contour of the slope. (C) Stakes are used to anchor the fascine bundles in place until the roots begin to grow. (D) On steep slopes, fasteners may be necessary to prevent bundles from being dislodged. (Smith, O'Connor, and Uchiyama, 1995)

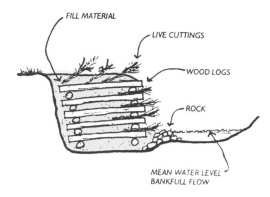

FILL MATERIAL

LIVE CUTTINGS

WOOD LOGS

ROCK

MEAN WATER LEVEL
BANKFULL FLOW

CROSS-SECTIONAL VIEW

Figure 63. Live crib walls are designed to be natural-looking retaining walls used to stabilize eroding stream banks. (Smith, O'Connor, and Uchiyama, 1995)

When complete, a revetment made with one row of overlapping trees has anchors with cable wrapped around the trunks of both trees on the ends. Also, the top of each of the middle trees is attached to the trunk of the previously laid tree with cable and the cable is then anchored to the bank. A revetment with more than one row is made of sets with each set of trees anchored and cabled together as a unit up the bank slope. Each set has an anchor at the bottom (near the toe of the slope) and at the top (on the bank near the top tree in the layer). Cable is threaded through both anchors and wrapped around each trunk in the stack of trees. If more than one set of trees is needed, each additional set of trees is placed overlapping with the previous set. (Figure 64). Use revetments on medium to small streams where there are moderate amounts of erosion, the water depth at the toe of the slope is less than three feet, and when the use of large machinery or the grading back of the slope is not possible. Revetments are economical strategies that do not require moving much soil and can be installed by several people in good physical condition. This enhancement method is best used where evergreen trees are plentiful. Freshly cut cedars, firs or pines work best. Recycled Christmas trees can be used if they are at least 10 feet tall.

Structural Techniques

Rock Riprap

Rock riprap is a method of placing stones or crushed rock along the stream bank to stabilize soil. Rock riprap is often applied to stream bank locations that encounter fast-moving water and are therefore most susceptible to erosion. Riprap is also a common technique to secure the toe (bottom) of a vegetative restoration site because it provides stability while plants become established (Figure 65). Be aware, however, that riprap has been challenged by many resource agencies concerned about how riprap can degrade wildlife habitat and increase sediment loads over time. It usually is not effective on steep slopes because rocks tumble into the stream during high water. It can be an expensive approach in large-scale projects if materials are not available nearby (trucking and labor costs must be considered). Also, riprap may cause problems by initiating erosion farther downstream.

There are two methods for installing rock riprap. One is to simply dump stones along the stream bank, letting the stone fall where it may. The other is called **imbricated riprap** and involves hand-placing large, angular stones close together. Although the first method is easier, it is not as attractive as hand-placed riprap and may not be as effective, as loose rocks can tumble into the stream or be washed away by high waters. In both cases, rocks must be installed during low water levels, they must extend to the bottom of the bank, and they must be large enough to stay in place during flooding.

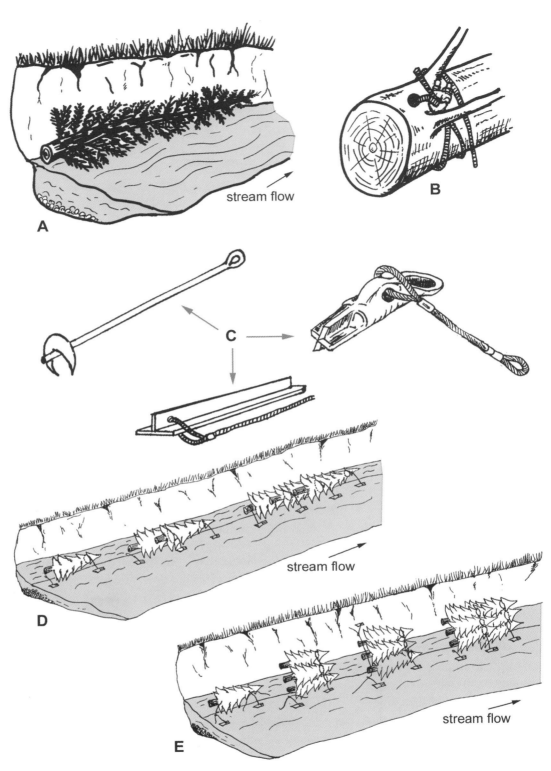

Figure 64. When installing a tree revetment, (A) place the trunk end upstream. (B) Wrap the cable around the trunk above the lower limbs or thread it through a hole drilled into the trunk and secure it with a clamp. (C) Disk anchors, T-posts, or duckbill earth anchors can be used to secure the cable to the bank. (D) Also anchor the top of the tree to the bank. If using more than one tree, work upstream and place additional trees overlapping each other. (E) When a one-row revetment is not adequate, additional rows may be installed on top of the first row to protect the bank. (Izaak Walton League of America figures)

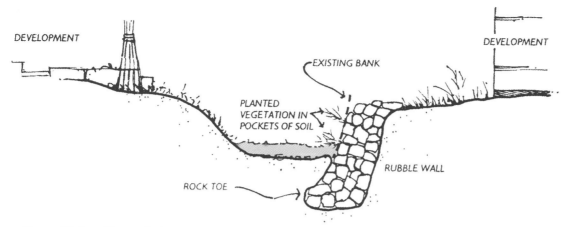

Figure 65. Installing rock riprap can protect newly planted riparian buffers from strong currents. (Smith, O'Connor, and Uchiyama, 1995)

The shape of the riprap always should fit in with that of the adjacent banks and should reach above the highest floodwater level so vegetation can be planted. Vegetation should be planted on top of the bank and in soil pockets between rocks to improve aesthetics and habitat quality. In very soft soils, a biodegradable, geotextile cloth can be installed beneath the rocks to provide a solid base.

Installing hand-placed riprap requires that the bank be sloped to a 2:1 ratio. A surface layer of stone is placed in an arrangement similar to a stone wall. The stones should be at the same angle as the bank slope, and the total thickness of the stone should be at least two feet or more depending on the velocity of the stream. The hand-placed stones should extend to the bottom of the bank and be on a solid foundation. The entire installation should be inserted at least three feet into the bank at each end of the wall to prevent the flow from scouring or loosening the rocks.

Dumped riprap should be placed onto the stream bank at a depth of three feet. Stone size is based on the velocity of the stream. The stones should be applied over a six-inch layer of gravel or biodegradable, geotextile cloth at a slope not exceed-

ing a 2:1 ratio. The largest rocks should be placed on the lower levels, and rough rocks should be used on the outside layer. Some hand labor may be necessary to arrange the rock, but the installation is handled primarily by heavy machinery.

Deflectors

Like rock riprap, deflectors are commonly used where bank materials are weak and water velocities are high. Deflectors are triangular, rock-filled structures attached to one side of the steam bank that propel water upstream. They are used to create pools, divert the flow of water from an eroding bank toward the middle of the channel, and guide a stream into a more meandering pattern. The size of the deflector and its angle to the stream bank will depend on water velocity, stream width, and depth. The basic design consists of a log triangle with an apex angle of 30 to 60 degrees.

Best suited for low-gradient, meandering streams where water levels do not fluctuate, deflectors are useful in a variety of situations to divert water from an eroding bank or, when placed on both sides of a stream bank, to deepen the stream. They also can be used to narrow a channel

and increase the velocity of flow. The design variables most used for deflectors are orientation angle, effective length, crest height, placement site, spacing between multiple deflectors, and construction materials.

The top log of a log frame deflector should be above water at normal flow. Steel pins are used to secure logs to each other and to anchor or pin logs into the stream bottom at four-foot intervals. Where the logs meet the bank, they are anchored by extending them six feet into the bank. A brace log is installed every 10 feet, and all upstream logs overlap the downstream log. The upstream angle should be no greater than 30 to 40 degrees. Stones are placed in the frame, with heavy machinery inserting the bottom layer 12 to 24 inches into the streambed.

More than one deflector may be required; if so, the first deflector should be constructed upstream of an existing or potential erosion point. After installation, the deflectors should be tested using a float to determine if the current directs the float to another bank downstream; if it does, another deflector is needed. However, as more deflectors are constructed, the natural meanders of the stream are altered.

Although deflectors provide an economical approach to erosion control when materials are locally available, correct placement requires professional assistance. Determining the flow characteristics of the stream is critical to identifying the proper location for the deflector. If placed improperly, deflectors can increase erosion and cause collapse of the opposite bank. In addition, an incorrect angle or length of a deflector can cause problems downstream. Deflectors also can cause erosion on their own stream bank if the bank is not protected, by causing reverse

upstream eddies behind the structure. In addition, a bank log or boulders may be required on the opposite bank to prevent a new cycle of erosion.

Gabions

Gabions are wire boxes filled with soil and rock, and planted with trees, shrubs, and grasses, that are buried into an excavated stream bank. They are best used in an urban setting where development is close to the stream, near bridge supports and stream access sites (Figure 66). Creating and installing gabions is a labor-intensive process. Although volunteers can create gabions, installation requires heavy machinery.

Gabions should be installed during the stream's low water periods, with live cuttings added in the dormant season. As with **cribwalls** and rock riprap, gabions must be installed into the streambed and bank to prevent undermining by currents. If installed incorrectly, the stream bank will erode and the gabions wash away. Over time, the wire from gabions might corrode and fall apart. Another criticism of gabions is that fish gills can get caught in the wire, causing harm or death to fish. Properly installed gabions usually are covered with soil. In addition, trees can be planted inside gabions to achieve a more natural look.

Removing Obstacles

Obstacles can deflect a stream's current onto its banks, thereby increasing erosion. They also can cause a debris jam that results in flash-flooding later (Figure 67). Some obstacles, such as fallen trees, can be beneficial because they provide habitat to fish and other aquatic wildlife. If obstacles or piles of debris are causing problems, such as reduced channel capacity,

1. TRANSPORT IN COLLAPSED FORM

2. CONSTRUCT INTO BOXES, WIRING DIAPHRAGMS INTO THE CENTERS TO ADD TO THEIR STRUCTURAL STRENGTH

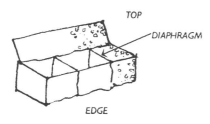

TOP

DIAPHRAGM

EDGE

A. EXCAVATE BANK BACK AND BELOW STREAM BED

B. PLACE GABIONS IN SLOPE USING STEPS (THE WIRE BASKETS)

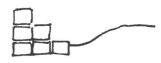

C. AFTER PLACING GABION WIRE BOXES, FILL WITH SOIL AND ROCK. TO FILL, PLACE LAYER OF ROCK ON THE BOTTOM AND FILL SPACES BETWEEN ROCK WITH SOIL. PLACE ANOTHER LAYER OF ROCK ON SOIL AND FILL IN WITH SOIL. CONTINUE THE ROCK-SOIL LAYERING UNTIL THE BASKET IS FULL. WIRE THE TOP DOWN.

SOIL
ROCK

D. PLACE SOIL ON TOP OF GABIONS

E. PLANT ON TOP OF GABION WITH GRASSES, SHRUBS AND TREES

3. CORRECTLY POSITION THE GABIONS INTO THE EXCAVATED STREAMBANK

EXISTING

WRONG WAY

EXISTING

EXCAVATE CHANNEL BANK

RIGHT WAY

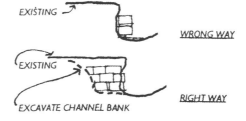

Figure 66. Gabions are typically installed along streams near which roads and urban development are located. (Smith, O'Connor, and Uchiyama, 1995)

care must be taken to address the problem without further harming the stream. If in doubt, leave the obstacle in place. Obstacles can be removed by hand, or by tractors, chainsaws, and other heavy equipment.

Using Multiple Techniques

A project may require use of more than one technique or a combination of vegetative, bioengineering, and structural techniques. For example, rock riprap can secure the toe of a bank while fascines stabilize the slope and plants in containers stabilize the top of the bank.

Controlling Erosion

Disturbing a stream bank to install an enhancement technique may increase the potential for soil erosion. It is important to create an erosion control plan that prevents downstream damage during installation and while plants are becoming established. For example, a silt fence at the toe of the bank will collect sediment until plants are established. A silt fence is a temporary structure to prevent or minimize transport of sediment in stormwater run off. The linear filter barrier is constructed of synthetic filter fabric, posts, and, depending upon the strength of the fabric used, wire fence for support.

Figure 67. Remove logs or debris that deflect flows toward the bank, back up flood flows, or excessively slow flows. (Izaak Walton League of America photograph)

have been detected through monitoring and fixed through maintenance.

Monitoring the Site

Most enhancement projects do not show immediate results. It might take several years to determine a project's success. Thus, monitoring protocols must be designed with the long term in mind.

In addition to designing a monitoring program based on information in Chapter Three, monitors can look for project maintenance needs. Regardless of the method chosen to stabilize a stream bank, observe the stream regularly after installation to see how it adapts to the project (Figure 69). Take photographs before, during, and after project implementation to document changes or improvements in the stream. The stream may not respond to the enhancement efforts as expected, especially during the first year. Note the way the stream adapts to seasonal changes and when its flow rates and water levels vary. Check the site during heavy rains and

The planted slope may also need to be covered with erosion control fabric. Natural fiber materials, such as coconut fiber or jute, are more desirable than plastic erosion control fabric for slope stabilization. The natural fibers will break down as the plants become established over several growing seasons (Figure 68). A local erosion control inspector might be available to help design an erosion control plan and get the plan approved before beginning work.

Providing for Ongoing Project Monitoring and Aftercare

Monitoring and aftercare are integral to the project's success. Unfortunately, monitoring and maintenance are parts of the project that usually are ignored. It is important to include monitoring and maintenance in the enhancement plan and to identify the people responsible for these tasks. Most projects fail because of minor problems that could

Figure 68. Covering newly planted slopes with erosion control fabric prevents soil loss until vegetation is established. (Izaak Walton League of America photograph)

periods of high flow. How are the different techniques used performing under different seasonal influences? Are the structures doing their job or are alterations necessary?

Monitor the project for at least one year following installation to check for insect infestation, excessive weed growth, dead plants that need to be replaced, erosion or gully formation, and other problems. Continue conducting biological monitoring protocols to determine if water quality at the site has improved. If silt fences have been installed, make sure they have not fallen down. Additional stream bank enhancement projects might be needed to adjust to changes in surrounding land uses and natural disturbances. For example, a year of increased development in the watershed or heavy rainfall may require securing the stream bank toe with additional riprap to prevent erosion.

Figure 69. At the end of a two-year period, 50 percent or more of the original plantings should be healthy and growing well. (Federal Interagency Stream Restoration Working Group, 1998)

Aftercare of the Site

In addition to fixing problems identified through monitoring, aftercare may involve reinstallation of portions of the project after storms; protecting plantings from wildlife and vandalism; watering, weeding, and mulching plants; and educating the surrounding community to prevent property managers or helpful neighbors from mowing or cutting down the project.

Be sure livestock and other mammals do not destroy plants before they have a chance to take hold. Repellents for beavers, muskrats, and deer are available commercially. In addition, aggressive seeds may compete with stream bank stabilizing vegetation for sunlight and nutrients. Weeding may be necessary, especially at the beginning of the growing season. Over the course of years, some plants will die of natural causes and provide more room for others. If high plant mortality occurs, an assessment should be made and the area replanted with appropriate native species. For more details on protecting plants from damage by wildlife, see Appendix N.

Tip Box

Vegetation may take more than one growing season to become fully established. The following examples provide a gauge of success.

Plant Growth after Two Years

Live stakes	70 – 100%
Live fascines	20 – 50%
Live crib wall	30 – 60%
Branch packing	40 – 70%

Table 5. Estimated Costs of Various Bioengineering Methods.

TECHNIQUE	COST
Pole cuttings	The cost will be for labor to prepare and plant the stakes, plus for any equipment that needs to be purchased. The estimated cost per linear foot ranges from $20 to $30.
Branch packing	Depending on labor costs for obtaining dead plant material and live pole cuttings, the cost can range from $1 to $10 per linear foot. The only material needed for installation is wire.
Brush layering	The cost ranges from $15 to $80 per layer per linear foot. Cost of labor, use of geotextile erosion control fabric, and use of machinery to grade back slope will increase the cost.
Fascine bundles	The labor to install fascines is the greatest cost. Plant cuttings, if found close to the site, can usually be acquired at no cost. Other materials include wooden stakes, twine, and, if needed, geotextile fabric. The estimated cost ranges from $1.50 to $20.00 per linear foot per fascine row.
Tree revetments	The cost will be approximately $20 per linear foot, including labor.
Rock riprap	This can vary depending on the labor, the height and extent of the (hand laid rock) wall, and if the rock materials are purchased or recycled. The range is approximately $75 to $200 per linear foot.
Live crib wall	Depending on the source of the wood logs and the height of the wall, a crib wall could range in price from $250 to $750 or more per linear foot.
Gabions	The cost ranges from $600 to $800 per gabion basket installed. Each basket is three cubic feet in volume.

Bioengineering projects provide major maintenance advantages over structural projects. After an initial maintenance period, bioengineering projects will maintain themselves as plant roots stabilize. Structural projects can weaken and deteriorate over time. This makes bioengineering projects much less expensive to maintain than structural engineering projects (Table 5). In addition, volunteers can maintain and monitor bioengineering projects.

In some cases, projects may need to be revised to better respond to changes in the stream or its watershed (Table 6). Remember, streams are dynamic systems in constant motion. The success of a project will depend on careful planning, help from the project's technical team, and regular, long-term maintenance. Time, effort, and commitment are needed to improve the health of a waterway.

Considerations when developing a monitoring and aftercare plan:

• Time/season to implement vegetation;

• Plant mortality and replacement;

• Wildlife management;

• Proactive weed control;

• Irrigation;

• Storm management considerations;

• Maintenance schedule; and

• Technical assistance.

Table 6. Potential Effects of Selected Stream Enhancement and Watershed Management Practices on Water Quality.[1]

Restoration Activities	Fine Sediment Loads	Water Temperature	Salinity	pH	Dissolved Oxygen	Nutrients	Toxics
Reduction of land-disturbing activities	Decrease	Decrease	Decrease	Increase/decrease	Increase	Decrease	Decrease
Limit impervious surface area in the watershed	Decrease	Decrease	Negligible effect	Increase	Increase	Decrease	Decrease
Restore riparian vegetation	Decrease	Decrease	Decrease	Decrease	Increase	Decrease	Decrease
Restore wetlands	Decrease	Increase/decrease	Increase/decrease	Increase/decrease	Decrease	Increase	Increase
Stabilize channel and restore under-cut banks	Decrease	Decrease	Decrease	Decrease	Increase	Decrease	Negligible effect
Create drop structures	Increase	Negligible effect	Negligible effect	Increase/decrease	Increase	Negligible effect	Decrease
Re-establish riffle substrate	Negligible effect	Negligible effect	Negligible effect	Increase/decrease	Increase	Negligible effect	Negligible effect

1 This table is based on information in Federal Interagency Stream Restoration Working Group, 1998.

Chapter Five

Stream Bank Enhancements on Range and Pasturelands

*A*merica's earliest grazing operations viewed streams solely as water sources for livestock. Although the banks and adjacent areas were recognized as providing excellent grazing and shelter for cattle, there was little understanding of the interactions among water, soil, vegetation, and grazing livestock. Traditional grazing strategies were generally designed for uplands, while riparian areas were overgrazed.

If allowed continuous access, livestock spend most of their time in close proximity to the riparian areas. Heavy livestock grazing degrades stream corridors through trampling, excessive vegetation removal, and manure and urine contamination. Livestock trampling compacts the soil, reduces water infiltration, and increases storm runoff and peak flow discharges. Stream flow becomes more

Figure 70. Newly installed stream bank fencing and riparian trees protect this stretch of Mill Creek, Pennsylvania, from livestock damage. (Izaak Walton League of America photograph)

variable and increased erosion may decrease bank stability and shear away portions of the stream bank. Stream channels become wider and shallower and gravel bottoms become covered with finer sediment, resulting in poor fish spawning habitat.

During the last few decades, government agencies, livestock producers, universities, and others linked to livestock production have developed an impressive body of scientific work regarding grazing management and natural resource enhancement (Figure 70). This chapter summarizes some of the most pertinent and practical information from these resources. A further reference on rangeland ecology, grazing, and riparian enhancement and protection techniques is located in the Watershed Stewardship Resources document available on-line at *www.iwla.org/ sos/resources*.

This chapter discusses the challenges associated with livestock grazing of riparian vegetation and outlines sustainable, economically viable resource management strategies for livestock production and a holistic, interdisciplinary approach to implement these efforts (Figure 71).

History of Rangeland Management

Many activities have affected range and pasture lands during the past 100 years, including overgrazing, forestry operations, and wildfire suppression. The active suppression of wildfires is perhaps the most important of these contributing practices. Wildfires are natural phenomena that once were viewed as wasteful and detrimental to natural resources. Wildfire suppression prevented the natural course of ecological **succession** in many areas, drastically altering the vegetative patterns in undesirable ways. Wildfire suppression policies of the past were grounded largely upon incorrect perceptions about the ability of the land to rebound from this natural disturbance. Today, resource managers understand the importance of wildfire as an integral part of the regeneration and renewal process in many ecosystems.

Figure 71. Rangeland management is a sound practice — ecologically and economically — which can also have aesthetic and recreational rewards. With cattle secured behind a fence, an angler casts his line in clean water at Milesnick Ranch in Belgrade, Montana. (National Cattlemen's Beef Association photograph)

Prior to the mid-1930s, grazing on the vast, largely unsettled public lands was not regulated. Livestock, sheep, and their herders wandered the open range looking for forage wherever it could be found. Herds often spread over large areas during their wandering and were gathered together when it was time to go to market.

The numbers of livestock steadily increased every year in order to feed the growing nation. There were no strict property lines, although individual livestock producers attempted to fence off parts of the range for their private use; and there were no incentives for livestock producers to use range resources in a sustainable manner. It did not make sense to invest in range improvements or to curtail damaging practices when other livestock producers in the common area could reap the rewards.

At the beginning of the 20th Century, the trend in deteriorating range condition was undeniable. No one anticipated the effects of uncontrolled grazing on the seemingly endless range resource (Figure 72). As a result, millions of acres of

Figure 72. The effects of poor agricultural practices, combined with years of sustained drought, caused the 1930s Dust Bowl in the Southern Plains. (Boettcher, et al., 1998)

rangelands produced forage below their potential. Livestock producers recognized the problems but saw no reason to conserve a resource that was available to everyone, yet the responsibility of no one.

Congress responded by passing several homestead laws intended to move unregulated public rangelands into private ownership. One hundred million acres were transferred to private owners. The more productive areas with readily available water were settled as base properties for livestock production businesses. On these lands, moist riparian soils supported vegetable and grain crops. More importantly, productive riparian areas were used to produce hay, a valuable source of food for livestock in the winter. This transfer of lands and subsequent establishment of more centralized operations gave birth to the private ranching system.

As early as 1891, the federal government allowed grazing under a permit system in national forest reserves. These permits did little more than designate who was allowed to graze in which part of the forest. In 1900, the federal government published permitting rules that were based on the number of animals the forest could support and still meet the forest reserves' purpose of providing timber and a continuous flow of water. In 1905, despite vehe-

ment protests from livestock producers, a grazing fee system was initiated. Permit holders quickly realized the benefits of grazing under the government restrictions, not the least of which was the guaranteed grazing privilege itself. In a 1915 speech, Secretary of Agriculture David E. Houston acknowledged that livestock producers were demanding a similar permit system on the remaining unreserved public lands.

World War I hindered improvements to grazing practices on public lands. As the demand for beef skyrocketed, livestock numbers on forest reserves and other public lands increased by 25 percent despite obvious and lingering signs of decades of overgrazing. The war necessitated increased beef production, and range managers were pressured to halt their range improvement programs, the majority of which centered on reduced stocking rates. Increased stocking, combined with droughts, disease, and grasshopper outbreaks, exacerbated range decline. Improvements made by some land managers were reversed completely by the combined effects of the war and natural disasters.

Ranching, as we know it today — characterized by a privately held property — became more popular with livestock producers in the 1920s. Producers who acquired a privately held base property could invest in range improvement and maximization of forage production. Those who owned traditional, free-ranging herds noticed the advantages enjoyed by their ranching colleagues and slowly began to switch to ranch-style operations. Because the range was so damaged and continued to be overgrazed, producers relied more heavily on hay and other supplemental feeds and development of water-holding structures and catchments. They also parceled the range into smaller, more intensively managed units (pastures). Producers gradually were forced to become better land managers or go out of business. Several ranchers achieved small victories in range improvement on their private holdings. Many built water-retention structures to provide water away from sensitive areas and planted hay and other forage crops to restore degraded lands (Figure 73).

The problems associated with unrestricted grazing are not confined to the history of the West. The problems described above also were recorded in the East during the early 1800s and in the South soon thereafter. Woodland grazing of livestock caused soil compaction and increased runoff and sedimentation of woodland streams. When livestock were turned into forests denuded by timber harvesting, they uprooted tree seedlings and severely protracted forest recovery. Many cleared forests were burned regularly to maximize the growth of grasses and **forbs** (leafy plants that are neither woody nor grass-like) for livestock consumption. This was true particularly in the southern pine forests. As the livestock industry pushed westward into the frontier, the East experienced less grazing pressure. With the invention of the cotton gin and other agricultural innovations, many southern pastures were converted to more profitable use as croplands.

Types of Range and Pasturelands

For many people, the term "range" evokes images of cowboys herding seas of longhorn cattle amidst vast expanses of cacti and tumbleweeds in the old West. This stereotype could not be more misleading. In truth, rangelands encompass many diverse ecosystems, from the pine forests of Alabama and the deserts of New Mexico to the Dakota grasslands and the high desert of eastern Oregon. Rangelands include shrub lands, tundra, coastal marshes, and any other ecosystem that supports grasses, grasslike plants, or other types of vegetation upon which livestock graze.

Pasturelands are rangelands characterized by actively cultivated, high-quality forage species such as alfalfa or fescue. These areas are managed to maximize forage production. Woodlands, another type of rangeland, are forested areas where grazing occurs. Rangelands in the East are predominantly pasturelands and woodlands.

Professional range managers and scientists discuss rangelands in terms of their characteristic vegetative communities. This is necessary because different

Figure 73. Hay farming, such as this operation at Milesnick Ranch, provides a less destructive alternative to allowing cattle free-range access to public lands. (National Cattlemen's Beef Association photograph)

Figure 74. This pasture from Milesnick Ranch is actively managed to maximize production of high-quality cattle forage. (National Cattlemen's Beef Association photograph)

vegetative communities must be cared for in different ways to achieve desired management objectives. In the interest of simplicity and clarity, this chapter will lump all of these lands under the term "rangelands."

Throughout this chapter, pastureland and pastures are distinguished differently. Pastures are individually fenced units of rangeland (Figure 74). A more descriptive name for pastures is "management units." The importance of pastures in effective riparian management and enhancement is discussed throughout this chapter.

Rangeland Ownership in the United States

Rangelands are publicly and privately owned. Private landowners who either use the land for their own grazing operations or lease the land to other livestock producers or other interests hold private rangelands. Public rangelands — those managed and administered by federal and state agencies — are divided into allotments. Private livestock producers pay a grazing fee for the privilege of using the public rangeland resources for their livestock operations.

Public lands used for grazing are administered by several federal agencies, including the Department of Defense, the Bureau of Indian Affairs, and the Fish and Wildlife Service. However, the Bureau of Land Management and the Forest Service administer the vast majority of federally controlled public grazing land.

In the West, livestock producers use both private and public lands for their operations. Eastern livestock producers operate almost exclusively on private lands. In addition, Eastern livestock producers tend to work on pasturelands or woodlands, and their operations are typically characterized by high animal density and small-pasture systems. The cultivated forage on pastureland and the abundance of vegetation in humid Eastern forests facilitates such a system. In the West, there is far less rain and far less grazing on pasturelands and woodlands. The arid climate limits the availability of forage from native vegetative communities. To compensate, operations have comparatively low density (when compared to Eastern livestock operations) because livestock require more land per animal to meet basic consumption needs (Figure 75).

Livestock Use of Buffer Areas

Understanding why livestock congregate in riparian areas is necessary before planning a riparian enhancement project. Many techniques presented in this chapter attempt to control livestock use of riparian areas. In some cases, attempts to lure livestock away from riparian areas will not be sufficient; structural approaches may be necessary.

The availability of drinking water is the foremost reason livestock congregate in riparian areas.

Another important reason is the abundance of high-quality, highly palatable forage. The constant presence of water promotes the growth and persistence of vegetation, including grasses, grass-like plants, and other edible vegetative species in marshes, meadows, and streamside woodlands. In many areas, upland forage species are only seasonally available because they dry up and lose their attractiveness during hot summers. Unless degraded by seasonal and other local conditions, riparian species show consistent new growth and maintain abundant forage.

Forage availability in riparian areas is very important. For example, a Montana study published by the Montana Agricultural Experiment Station showed that during August and September, livestock received 80 percent of their forage from riparian areas, even though the riparian area constituted only four percent of the total pasture acreage. Livestock linger in riparian areas and may not graze new growth on upland vegetation even when riparian vegetation is severely overgrazed.

Livestock also tend to congregate in riparian areas for shelter. Healthy

woody vegetation provides shade during hot summers and excellent cover against rain, wind, and snow. Unfortunately, livestock can severely damage these woody species by over browsing the trees and shrubs. If overgrazing prevents new growth and replacement, desirable woody vegetation eventually is eliminated (Figure 76).

Another reason livestock tend to spend time in riparian areas is the position of these areas in the landscape. Cattle

Figure 76. Livestock that are allowed to range freely can have detrimental effects on stream banks, riparian vegetation, and water quality. (Izaak Walton League of America photograph)

prefer areas with relatively low slope. In fact, as the slope of the land increases, cattle use decreases. Cattle tend to graze flat areas before moving to comparatively challenging slopes. Slope needs to be considered when planning a project to lure livestock away from riparian areas.

Overgrazing in Buffer Areas

Through sound management, it is possible to graze riparian areas and still maintain or enhance their condition. Many livestock producers have demonstrated that grazing can occur without harming the land. The challenges presented below are associated with continuous, unrestricted grazing.

Riparian area degradation begins when the rate of forage consumption exceeds or impairs the new growth and reproduction of natural forage. In situations where the rate of forage consumption exceeds the rate of forage regeneration, the riparian area will not be able to support that level of grazing. Under these conditions, streams cannot adequately filter sediment, recharge aquifers, or slow floodwaters. These changes harm fish and wildlife by reducing the quality of forage, water, and shelter. Livestock will suffer as well (Figure 77).

In addition, repeated overgrazing will limit the availability of native forage. This can lead to higher operating costs (in the form of supplemental forage) to maintain animal weights and prevent malnutrition. The reduction in groundcover increases the susceptibility of the soil to erosion. Runoff or direct deposition of manure into streams can lead to excess nutrients in the water, causing algal blooms and reduced oxygen levels. Waterborne disease transmission also may occur, particularly in high-animal-density operations.

Livestock trample and compact soil, gradually driving out the soil-binding perennial vegetation and replacing it with shallow-rooted species. These woody perennials are unable to naturalize when livestock consistently consume or trample delicate saplings and compact the soil.

Soil compaction also prevents deep infiltration of rain or snow into the soil. Groundwater tables are not recharged, and the amount of surface runoff increases. Eventually, increased runoff and the absence of soil-binding vegetation lead to sedimentation of waterways. Habitat for fish and other aquatic life is degraded.

Soil compaction can occur even if the riparian area is not overgrazed. The soil's susceptibility to compaction depends on

Figure 77. In an overgrazed system, vegetation is consumed faster than it can naturally regenerate. Two plots of rangeland pictured here from Mortenson Ranch in Hayes, South Dakota, show the difference between (A) an overgrazed and (B) an actively managed plot. Appropriately managed livestock operations do not degrade riparian areas, and can allow the growth of vegetation for forage as well as for stream bank protection. (Boettcher, et al., 1998)

many variables, two of which are soil composition and the timing and distribution of livestock grazing. Soil composition is a natural factor that must be considered in any riparian enhancement program. Implementing an appropriate grazing system can control the timing and distribution of livestock grazing. Grazing systems will be highlighted later in this chapter.

The loss of wildlife resources is one of the most significant effects of overgrazing in riparian areas. This is particularly true in the West, where water is confined to small, well-defined areas of the landscape. Although riparian areas account for less than one percent of Western lands, they are the most productive areas for wildlife. Unlike generally uniform upland regions, riparian areas are diverse in the number and types of available habitats. This diversity supports an abundance of wildlife.

Wildlife relies on riparian areas as travel corridors and readily available supplies of water, cover, and food. More bird species are found in riparian areas than in all other western vegetation types combined. In Arizona and New Mexico, nearly 80 percent of all amphibians, reptiles, birds, and mammals depend on riparian areas during some portion of their life cycle. Forty percent of these animal species are completely dependent on riparian areas. Studies in Oregon and Wyoming have documented similar dependency. A Pennsylvania State University study in Lancaster County, Pennsylvania, documented that the number of nesting attempts made by birds nearly doubled after livestock were fenced from riparian areas.

Chapter One reviews the basic physical characteristics of a stream. Overgrazing can change the size, shape, and flow of a stream, which affect other stream qualities and uses.

The Status of Riparian Areas on Rangelands

Much of the disparity between upland and riparian health stems from a historical lack of knowledge concerning riparian ecology and a lack of riparian area management techniques. Scientific tools for predicting the effects of grazing and formulating management and recovery strategies still are quite new. In the late 1800s and early 1900s, ranchers and land managers had no way to predict the effects of grazing before the damage was done.

The last 40 years have witnessed a great deal of effort to better understand the impacts of livestock grazing on the function of riparian areas. Most recently, the challenge moved from understanding these effects to creating management tools that minimize these impacts. Rudimentary riparian recovery tools now exist, and many livestock producers are implementing riparian management initiatives. Most of these initiatives are voluntary, implemented with the objective of achieving increased animal yield and higher profits. Riparian area management is a sound practice — ecologically and economically — which also has financial rewards that can quickly outweigh initial investments.

Responsible Grazing Management is Good Business

The degradation of riparian areas translates into an enormous economic loss to livestock producers. The reduction in the natural forage base means increased feeding costs, especially during the winter months. Most importantly, potential animal production levels may not be realized. Disease, poor forage, and limited water availability are all economic losses that could potentially be traced to deteriorated riparian areas.

For private landowners, responsible grazing management may mean an increase in their property values. Besides lower productivity, degraded aesthetic values are another obvious effect of overgrazing on property values. Overgrazed lands do not exhibit vitality, productivity, or any characteristics of an attractive investment. The loss of productive topsoil also reduces the land's value and the livestock producer's ability to produce hay and other alternative forages. Without nutrient-rich topsoil, the enhancement of native plant communities is compromised as well.

Livestock producers are not the only people who feel the effects of overgrazing on riparian areas. People far removed from the livestock operation may experience how overgrazing contributes to sediment loads in rivers and smaller streams and reduced water supply. Throughout time, heavy soil erosion can fill reservoirs and clog intake lines for irrigation and hydroelectric power facilities. Similarly, many regional economies depend on readily available water for irrigation of cash crops. Municipal reservoirs may require frequent sediment dredging. The costs of repairs and maintenance are borne by consumers in the form of higher utility bills and/or food prices. Similarly, the effects of grazing on the destruction of migratory fish and waterfowl habitat can have wide-ranging impacts on areas dependent on income generated through hunting and fishing.

Enhancing degraded riparian areas and implementing sound livestock management techniques may create new sources of income. Fees collected from outdoor recreationists and riparian timber production are two possible sources. Many ranchers can diversify their income base

by charging entrance fees to recreationists or by running hunting/fishing lodges (Figure 78). Riparian forestry has been practiced widely in the East and is growing in the West. Riparian forestry requires little input in terms of labor or capital and offers an easy opportunity for additional income. State forestry agencies may be able to provide additional information on riparian forestry programs.

Developing a Rangeland Management Plan for Stream Enhancement

As stated in earlier chapters, it is important to consider the entire watershed

Figure 78. Careful rangeland management can generate financial profits from hunting and fishing activities that outweigh initial investments. This healthy riparian habitat at Dave Wood Ranches in Coalinga, California, supports both grazing for cattle and opportunities for fishing. (National Cattlemen's Beef Association photograph)

when developing a rangeland management plan for stream enhancement. The condition of upland resources must be considered along with upstream land uses, which might affect the intensity, volume, and quality of stream flow. Uplands must be managed to maximize forage production at the same time riparian areas are being enhanced; otherwise, stream recovery may be impossible.

A riparian enhancement project should be the product of an interdisciplinary technical team with different resource management specialties. A typical technical team should consist of the property owner, a botanist, a wildlife or fisheries biologist, a hydrologist, a grazing management specialist, and, especially on public lands, members of the public with an interest in the health of the riparian area. The importance of a team approach cannot be overemphasized. Interdisciplinary teams can more effectively identify the key management considerations at a specific site and more thoroughly manage all ecosystem characteristics for maximum results. Appendix B includes information on how to assemble a technical team.

There are no simple solutions to riparian enhancement, and each enhancement project must deal with site-specific issues. The methods presented are only suggestions and will not always apply to every livestock operation. The principles, techniques, and recommendations in this chapter draw information from case studies in many vastly different regions of the country.

The most important thing to remember when creating a riparian enhancement program is to control the timing, duration, and intensity of grazing. The methods presented in this chapter can be used to control these three critical factors. Eight

basic grazing models are presented, followed by discussions of structural improvements (fencing, water facilities, etc.) that can be used to ensure that the objectives of the grazing management plan are realized.

Remember to not expect results overnight. Projects that set goals for long-term recovery are the most successful. Riparian enhancement projects may show dramatic responses in a few years or a few decades, depending on the level of degradation. Where overgrazing and loss of topsoil have been particularly severe, it may even be necessary to begin the enhancement project with the reintroduction of native riparian vegetation.

Grazing Systems

Much research has been devoted to developing and monitoring the effectiveness of different grazing systems. Many livestock producers are reluctant to adopt grazing systems, thinking that reduced access to higher quality and quantity riparian forage will adversely affect the weight gain and subsequent marketability of their livestock. However, livestock producers working under sustainable grazing management systems that control the season, duration, and intensity of riparian grazing have healthy herds that gain weight more quickly and suffer fewer incidents of disease.

Grazing systems are as varied as the landscapes and landforms on which livestock graze. Before discussing some common models for grazing systems, it is important to list some guidelines for developing a site-specific, customized grazing system. No matter what the goals of a grazing system, the goals of the system should attempt to address all

factors that contribute to a healthy range, not just a healthy riparian area. The effects of any proposed system on each of these factors should be considered carefully. Consideration of these factors will help identify areas particularly susceptible to damage and other areas that are more resilient because of their physical and biological attributes. See the tip box at left for examples of factors to consider.

When designing a grazing system, remember that livestock are not the only animals that use riparian areas for forage, water, and shelter. The amount of forage and seasonal patterns of forage consumption by wildlife must be considered as well. Wildlife use must be factored into grazing strategies where populations of large herbivores, such as elk, antelope, or deer, coexist with livestock in the riparian area. If wildlife use rates are not considered, the estimated amount of forage available to livestock may be higher than the amount actually available. Because woody species are so important to riparian health, it is particularly important that wildlife use of woody riparian species be monitored and accounted for in the riparian enhancement and protection program.

Other factors also are important to design an appropriate grazing system. Remember to consider the following:

- Biological needs of primary forage, woody, and other vegetative species;
- Tendency of soil to erode;
- Annual distribution of precipitation;
- Slope and topography;
- Type, age, and sex of animals;
- Forage used by big game;
- Requirements of non-game wildlife;
- Water quality and quantity;
- Stream bank and streambed protection;
- Present and potential ecological conditions.

Reducing the Total Number of Livestock on the Land

Reducing livestock numbers may not induce riparian recovery automatically. Proper stocking density is important, but merely reducing livestock numbers may do little if continuous, unrestricted grazing continues to be practiced. In fact, livestock numbers may not be the problem. The problem most likely lies in the timing and duration of grazing in a particular area.

Without proper control, livestock will spend a disproportionate amount of their time in the riparian area. Even with reduced numbers of livestock, riparian vegetation may not receive sufficient rest to maintain growth and vigor. Plants must have an opportunity to store energy during the growing season so they can sprout or leaf again the following spring. Energy is stored in the roots in the form of carbohydrates. If grazing continually removes photosynthetic structures such as leaves and green stems, plants cannot produce these carbohydrates. Eventually, the plants will no longer be able to replenish themselves.

Similarly, stream banks still will be trampled during periods of high soil moisture. Stubble heights and litter levels (vegetative residue from the previous year) may not be sufficient to slow and filter runoff or to protect stream banks. Livestock reduction can be considered an element of an overall riparian management and enhancement plan, but should not be considered a solution by itself.

> **Consult a local extension service, Bureau of Land Management, Forest Service, or Natural Resources Conservation Service field office for information about determining a proper stocking rate.**

The Land Through the Seasons

Seasonal changes in the biological productivity and hydrology of the watershed greatly influence management considerations. Seasonal changes in hydrology are site-specific and are not addressed, except in the most general terms. Seasonal differences in the growth and productivity of vegetation are summarized below.

In the discussion below, seasons are not defined strictly by assigning months to each season. Because the exact onset of seasonal changes varies with latitude, elevation, and the growth preferences of native plant species, seasons are described according to the changes that occur in soil moisture, plant growth, and other physical and biological factors. The precise time of the year when these changes occur should be determined for each project.

Winter

Winter is a time of little or no plant growth. Depending on upland conditions, livestock and wildlife will use riparian areas most heavily during the winter for forage and shelter. Woody plants are dormant, and vegetation has finished storing food for the spring. Frozen soils are more resistant to compaction. Stream bank soils are not as likely to be pushed into the stream by livestock activity. If grazing on the year's woody growth and the amount of stubble and litter is monitored closely, winter grazing can have relatively few effects. In some areas, snow cover may prevent grazing during the winter.

Spring

In many places, spring is characterized by flooding. Livestock will be forced onto the uplands at times, reducing the pressure on the riparian area. Upland vegetation begins to grow in early spring, providing

Figure 79. Summer is the primary growth season for most perennial grasses and woody vegetation. However, as summer heat dries out the uplands, riparian areas become more appealing to livestock. (Jerry N. McDonald photograph)

forage away from the riparian zone. Saturated soil in the riparian area is highly susceptible to livestock disturbance. Animals can easily uproot plants from wet soils.

Depending on the seasonal growth characteristics of the native plant community, limiting grazing to the spring may give vegetation the entire summer and early fall months to grow free of grazing pressures. As with winter, enough stubble and litter should be left to slow and filter runoff and to protect stream banks.

Summer

Summer is the primary growth season for most perennial grasses and woody vegetation (Figure 79). As uplands dry out, upland vegetation will become less palatable to livestock. This situation, when combined with summer heat, will drive animals to riparian areas. In areas with extremely hot, dry summers, this may be the season when livestock use riparian areas the most.

Woody species are particularly susceptible to summer grazing. They used a large portion of their stored energy to re-leaf and initiate new growth in the previous spring. Preventing them from replenishing their energy stores during the summer will severely compromise their growth in subsequent growing seasons. Grasses may exhibit a similar response. In

many ways, the profusion of summer growth masks what is actually a highly sensitive time for vegetation.

Fall

The most important consideration in fall is the fact that most plant growth stops. This is not true for warm season grasses, which are completing their growth and reproductive cycles. Woody species are less susceptible to damage because the main food-storage season has passed. Dormancy reduces the effects of grazing even more. Cool season grasses have developed mature seeds so replacement is not undermined. Some upland species may be active again, and livestock may want to avoid falling temperatures by taking advantage of warmer, southfacing uplands. Enough vegetation should remain on the ground to protect soil and reduce runoff during the following spring.

Grazing in any season has advantages and disadvantages. How can a landowner or manager create a sustainable grazing program? As with any activity that involves natural resource consumption, there are trade-offs. The less desirable side of these tradeoffs can be minimized by careful observation and monitoring of range conditions. As soon as conditions begin to deteriorate, managers can remove or reduce the grazing pressure. This may be expensive, but in the long run a few growing seasons invested in a grazing system will nurture a more stable, resilient, and sustainable resource.

Grazing System Models

Although much research needs to be done in this area, attempts have been made to predict the response of vegetation in different types of watersheds to different grazing systems. Stream gradient and sediment load will also affect the vegetative response to different grazing systems. Considering the natural conditions (e.g., gradient and sediment load) of the stream, the health of vegetation and the quality of stream banks under different grazing systems can provide some indication of what direction a riparian enhancement project should take.

No Grazing

No grazing means a moratorium on livestock grazing. This is an economically unacceptable alternative for livestock producers. Feeding costs alone would be exorbitant. However, complete rest for the riparian area may be necessary to jump-start recovery in severely degraded watersheds. The decision to remove all grazing pressure depends on climate, vegetation, soil, and other factors. Like all natural resource decisions, it must be made on the basis of site-specific goals and ecological constraints.

Dormant Season Grazing

Under this system, riparian grazing is allowed only in the dormant season, when plants have completed their active growth period. Vegetation is allowed to photosynthesize and store energy without interruption. Because dormancy coincides with winter, soils may be frozen and less susceptible to erosion. This system may require supplemental feeding away from the stream and supplemental water from wells or springs. Spring growth can compensate quickly for forage lost to consumption during the winter.

Early Growing Season Grazing

This system takes advantage of conditions that exist during the spring by grazing in the early growing season. Plants grazed during the early growing season have the remainder of the season to grow and store energy. Soil damage and stubble height can be controlled with careful monitoring and management. Some supplemental feed for livestock may be necessary before the generation of new growth, or if new growth is covered with mud from floods.

A potential disadvantage of early growing season grazing is that it can have a drastic effect on the total production for the entire year. If maximizing total forage production throughout the year is a primary management objective, this system might not be appropriate.

Late Growing Season Grazing

Grazing the riparian area during the late growing season takes advantage of mid- to late-summer activities in plant physiology. Most plants have grown and reserved energy for the majority of the season, although woody vegetation is still sensitive. Many grasses have produced mature seed heads to ensure adequate replacement in following years. In many areas of the West, the later grazing is delayed, the better for the vegetative community. This will not be true in areas where warm-season grasses are the dominant forage species. Warm-season grasses are most common in the arid Southwest and coastal states of the humid Southeast.

Table 8. Rest Rotation Cycles in a Three-Pasture System.

	REST ROTATION CYCLE 1			REST ROTATION CYCLE 2		
	Year 1	Year 2	Year 3	Year 4	Year 5	Year 6
Pasture 1	Rest	Graze	Graze	Rest	Graze	Graze
Pasture 2	Graze	Rest	Graze	Graze	Rest	Graze
Pasture 3	Graze	Graze	Rest	Graze	Graze	Rest

Rotation

In a simple rotation scheme, livestock are moved among different pastures so that no single pasture bears the brunt of an entire year of grazing. All pastures are used during some part of the year.

This system can be very effective if it is timed to maximize the advantages and minimize the disadvantages associated with grazing in each season. Impacts are localized and grazing is tightly controlled. A rotational system based on seasons is called a **deferred rotation system**. Rotational systems that allow an entire year of rest from grazing are called **rest rotation systems**. Both are discussed below.

Rest Rotation

A rest rotation system allows each pasture in the system at least one year of complete rest. Livestock must use forage in other pastures and/or receive supplemental feed. Rest rotation normally involves a minimum of three pastures (Table 8).

As with every grazing system, rest rotation has its limits. Rest rotation can be an excellent system for sedge/rush/grass riparian meadows. It has not been as successful at restoring woody vegetation on lands west of Wyoming. In addition, rest rotation has been somewhat successful at restoring cottonwood stands in the desert Southwest when grazing is limited to the winter.

Where it has been successful, rest rotation has improved forage production by allowing each pasture an entire year's rest every third year. The period of rest is beneficial because it allows all plants to grow, store energy, and reproduce with no disturbances. This growth helps stabilize soils through root binding and improves interception of rain and upland runoff. When livestock are introduced into a rested pasture, the rate and amount of forage consumption must be monitored carefully. There may be no long-term benefits to rest rotation if livestock graze in such a way that the pasture returns to pre-rest conditions.

Deferred Rotation

Deferred rotation allows grazing to occur after vegetation has reached a desired stage of development. For example, it may be desirable to delay grazing until after grasses fully develop seed heads. Alternatively, it may be to the livestock producer's advantage to delay grazing until woody vegetation finishes its growing season and becomes dormant. Criteria will change depending on site conditions, the desired vegetation community, and the constraints of the livestock operation.

Table 9. Deferred Rotational System with a Three-Part Season.

	Year 1			Year 2			Year 3		
	Early	Mid	Late	Early	Mid	Late	Early	Mid	Late
Pasture 1	Graze	Rest	Rest	Rest	Graze	Graze	Rest	Graze	Graze
Pasture 2	Rest	Graze	Graze	Graze	Rest	Rest	Rest	Graze	Graze
Pasture 3	Rest	Graze	Graze	Rest	Graze	Graze	Graze	Rest	Rest

In a deferred rotational system, one pasture is rested for most of the growing season, while the others are grazed. The rest period is rotated to each pasture in successive years. Usually, one pasture is grazed early in the year (spring) and then allowed to rest for the remainder of the year. Grazing continues on the pastures deferred during the spring. For operations that do not graze year-round, the phrase "grazing period" can be substituted for the word "year" (Table 9).

In some cases, the pasture is grazed early in two successive years. Each pasture is free of grazing for two successive years during the fall and winter season. The advantages of this system differ from the one-year deferred rotation system. The increased deferred period can be particularly beneficial for shrub species that depend on late season carbohydrate storage. Large woody vegetation benefits as well if there is sufficient forage from grasses and forbs. Livestock begins to consume woody vegetation when **herbaceous** plant consumption reaches 40 to 45 percent and increases as herbaceous plant-use increases. If woody vegetation is grazed heavily in two successive late seasons — the most critical growth and energy

storage period — plant vigor will be compromised severely. As with all systems that control spatial distribution of livestock, controlling the amount of time animals are allowed to graze each area is the key to success (Figure 80).

Because the deferred portion of this strategy is based on the growth stage and reproductive status of the vegetation, remember to consult the botanist or grazing management specialist on the project's technical team. In addition, herding will be necessary to control livestock movement, and it is very likely that hay and other feed will be required since some forage is unavailable during part of the year. Eventually, with proper control of livestock numbers and adherence to the grazing management plan, the ecosystem should begin to replenish itself

Figure 80. Controlled grazing prevents damage to the rangelands at Dave Wood Ranches. (National Cattlemen's Beef Association photograph)

Figure 81. This riparian area at Mortenson Ranch has been fenced to restore the riparian corridor. (Boettcher, et al., 1998)

and become sustainable. As with any grazing system, the intensity of grazing must be controlled so stream banks and other sensitive areas are not compromised by the removal of too much vegetation.

Corridor Fencing

Corridor fencing is a fencing design that places the stream and riparian area into a relatively narrow band, creating a well-defined riparian corridor. Long spans of corridor fencing create green ribbons of riparian vegetation along waterways. Livestock are grazed in the riparian corridor when riparian grazing is most appropriate. Fencing portions of the riparian corridor may be the best alternative for rapid improvement of degraded riparian areas with little vegetation (Figure 81).

Although corridor fencing may be the only option available in some situations, this design has several disadvantages. Livestock will walk the fence line, creating trails, and exert more pressure on the fence over time. When animals are herded into the riparian corridor, they may not have sufficient freedom of movement, causing distribution problems within the fenced area. In addition, fences do require additional funding to build, and maintenance

also needs to be factored into this cost, especially in areas where streams experience high-velocity floods. Another consideration is how wildlife will access the water source.

If a land owner or manager decides that fencing is the best approach for recovering a severely degraded riparian area, creative designs should be considered that would permit the area to be included in future grazing systems and pasture rotations. Riparian pastures are one example of how this may be accomplished.

Riparian Buffer Pasture

Riparian pastures may contain both upland and riparian vegetation and should be managed to maximize benefits to both livestock and the natural ecosystem. Riparian pastures can be grazed or rested depending on current riparian conditions. Therefore, the objective of riparian pastures is not to exclude livestock from the riparian areas, but to provide for closer management and control of their use. Fencing the riparian area (and possibly some adjacent upland) to tightly control livestock access is one of the easiest ways to limit grazing pressure.

In a riparian pasture system, upland grazing can continue. Some drinking facilities might be required if access to water is restricted. Ideally, livestock should be allowed to forage in the riparian area at a time most appropriate for growth and replenishment of upland vegetative species.

Riparian pastures can be used fairly frequently if livestock are allowed to graze only for short periods. In some cases, these periods may be as short as one week. The important factor is to keep the animals moving, so they never linger in the riparian pasture for too long. An experienced livestock herder who understands these concepts can be an invaluable asset to any riparian enhancement project. However, a variety of factors such as terrain, the size of the riparian area, and fence construction and maintenance costs may limit the practicality of a riparian pasture system.

Structural Improvements

Structural improvements are used to help achieve the objectives of a rangeland management plan. Always consult with an extension service, Bureau of Land Management, Forest Service, or Natural Resources Conservation Service field office for information about design and construction specifications. Cost-share programs may be available to help defray some construction expenses.

The federal government allows livestock owners to create structural improvements on public lands. The structure must be consistent with the allotment management plan and must fulfill a management need. A formal, written agreement, whether a legal permit or a memorandum of understanding, usually is required between the federal government and the livestock owner. Design and construction must conform to agency specifications, most of which are very similar among the agencies. Structures built on Forest Service and Bureau of Land Management lands are considered federal property, not the property of the livestock owner who builds it.

Despite these ownership rules, there is a strong incentive for livestock produc-ers to invest in structural improvements to the public lands their livestock graze. If the profits derived as a result of the improvement are greater than the cost of building the improvement, then it makes sense to build the improvement. Improvements might not pay for themselves in the first two or three years, but as with any investment, long-term results are most important. Whether to invest in structural improvements to enhance riparian areas and maximize forage production is a decision that should be made in close consultation with a professional range manager with experience in agricultural economics. Consult the local extension office to locate professionals with this expertise.

Fencing

Fencing is often a necessary grazing-system tool. Fencing cannot, however, control the timing of grazing. This must be controlled by actively moving the animals around the property. A herder might negate the need for fencing if he or she is successful at controlling animal distribution. Livestock operations have eliminated extensive fencing after the grazed lands healed to a sustainable level. These fence-free operations keep the animals moving throughout the range. A more hands-off approach requires fencing to achieve desired animal distribution.

The use of fencing is a must in many operations; especially for livestock owners working with a small pasture system. If pastures have been laid out purposefully to include waterways, fencing is necessary to stop livestock intrusions. In such cases, fencing can be combined with reinforced crossing points to limit disturbance to the stream, while still allowing use of most of the pasture. Alternately, fencing can be

configured to provide water gaps — breaks in the fence where livestock are given limited access to the stream.

An effective fencing system can be designed to fit each site. Fencing designs should consider topography and climate, as well as the cost, age, and size of livestock (Figure 82). Wildlife access to migratory routes and wintering grounds is particularly important. The types of fencing described below create minimal wildlife obstructions. The wildlife biologist on the project's technical team can help customize a fencing design to meet wildlife needs and riparian enhancement objectives at the same time.

One of the most daunting aspects of fence construction is the labor and funding required to complete the task. Volunteer labor can be found and should be used in conjunction with professional assistance for riparian enhancement projects. For example, the Public Lands Restoration Task Force of the Izaak Walton League's Oregon Division has used volunteers successfully for fence construction and maintenance projects. Among these volunteers have been members of Boy Scout troops who are looking for merit badge projects.

Specifications for fence design and construction are available from government and extension service field offices. Always consult with local agencies about proper fence construction specifications and the most appropriate fencing technologies for the property.

Suspension Fences

Suspension fences are best used on flat or gently rolling terrain. They can be used as interior fences, but should not be considered for perimeter fencing. Unlike traditional fence construction, the wire is not stapled or nailed to the posts; it is

---

Tip Box

Build a Fence!

- **Corner posts** are at the corner of the fence. They must be strong, sturdy, and well anchored. Half of the corner post should be below the surface of the ground. Weak corner posts can cause the entire system to fall inward.

- **Brace posts** are the four (or sometimes more) posts placed adjacent to the corner post (two on each side).

- **Brace bars** run between the brace posts and corner posts to help stabilize the corner section. Two brace posts with a brace bar forming an "H" also are used at regular intervals along the length of the fence, usually every quarter of a mile and near gates, to provide strength and prevent sag.

- **Line posts** are placed at regular intervals along the wire to hold the wire up, maintain wire spacing, and prevent sag and tension loss. Posts may be constructed from wood, steel, or rock baskets (called rock cribs) depending on which structures best achieve the desired function of the fence with regard to management objectives. Longevity of different post materials is an important consideration that should not be overlooked.

---

allowed a limited range of motion inside a semicircular clip. This design allows some give so the fence moves during animal contact, reducing incidents of fence breakage and entanglement of livestock (Figure 83). Because line posts are placed at 80- to 120-foot intervals, suspension

Figure 82. Effective fencing design, like this one at Dave Wood Ranches, needs to consider topography and climate, as well as the cost, age, and size of the livestock. (National Cattlemen's Beef Association photograph)

fences require only one quarter of the number of posts of traditional fences.

Let-Down Fences

Let-down fences are extremely flexible. The majority of the fence can literally be "let down" during livestock herding or wildlife migration. They are best used in areas that experience heavy logging or snowfall (Figure 84).

Long spans of let-down fence should be constructed in sections to accommodate abrupt changes in topography. Each individual section can be let down. For example, if a let-down fence must cross a low spot in the terrain, it is better to construct two separate sections that meet at the lowest point. It is nearly impossible

to maintain tension or structural integrity in a one-part, U-shaped fence; two spans that meet at the low point will perform the same function more efficiently.

With let-down fences, the wire is not attached permanently to the posts, but is attached to wooden stays that are held to the posts by wire loops. A let-down fence is a preferred type of fencing where there is a high probability of fence breakage or other damage.

High-Tensile Wire Fences

High-tensile wire was developed in New Zealand and Australia and is used successfully in the United States. High-tensile wire is smooth and narrow, so animals are less likely to get injured or tangled in it.

High-tensile wire is installed much like a barbed-wire fence. The corner posts must be made of extremely durable, sturdy material to withstand the stress of high-tensile wire. Corner posts should be anchored along both fence lines with buried boulders, gabion baskets, logs, or

Figure 83. Wire suspended inside a metal clip allows a suspension fence to move during animal contact. (Marcy, 1986)

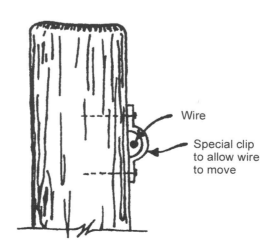

Wire

Special clip to allow wire to move

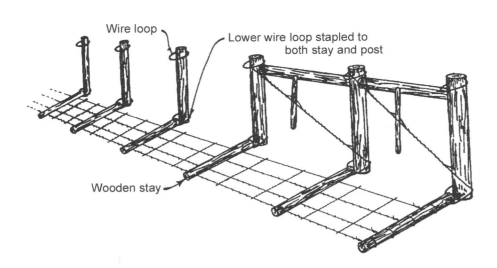

Wire loop

Lower wire loop stapled to both stay and post

Wooden stay

Figure 84. A let-down fence provides flexibility in areas that are designated for several uses, such as logging or wildlife migration. (Marcy, 1986)

other anchors to prevent leaning or uprooting. To add additional strength to the corner, the ends of brace bars must be chiseled to fit holes in the corner posts.

Monitoring tension in the wire is very important. Tension can be released during drifting snow or as other circumstances require. Seasonal changes can affect wire tension significantly. As the weather gets colder, the tension may need to be released because the wire will contract. Similarly, the wire will need to be tightened as weather warms and the wire expands. The fence can be tightened if it is made slack by a fallen tree, wildlife, or livestock.

High-tensile wire can be electrified with great success. The wire carries the current well, and electricity costs for high-tensile wire are lower than for conventional wire. Consult your range conservation district and fencing manufacturers for information about additional equipment, design requirements, and installation guidelines for constructing an electric fence.

Shelters/Buildings

One of the easiest, least-expensive ways to improve the riparian area is to lure livestock away from it. As stated earlier, livestock congregate in riparian areas for shelter, food, and water. Salt, feed supplements, scratch posts, and water should be placed in areas where livestock regularly travel. Creating desirable conditions on uplands can reduce grazing pressures on riparian areas and, if successful, may eliminate the need for fencing.

One way to lure livestock away from riparian areas is to build shelters or other structures that provide shade in the summer and shelter from winter storms. Shelters can be built out of readily available materials at minimal cost without much skilled labor. A well built shelter can last more than 30 years. Simple pole structures are the least expensive to build (Figure 85). Stud-wall sheds are appropriate shelter from harsh winter winds. Because young animals are particularly susceptible to cold-related illnesses, winter shelter provides additional health benefits.

Local extension offices can provide specifications for shelter construction and design. Some important principles are worth mentioning, particularly where site selection is concerned. Never place shelters in depressions that collect runoff or where snow tends to drift. Because

Figure 85. A pole shelter is an easy and inexpensive way to lure livestock away from the riparian corridor. (Izaak Walton League of America figure)

manure will collect in shelters, place them in such a way that prevents manure-laden runoff from reaching nearby water bodies. Sheds should have enough space to accommodate several animals. Summer shelters should be placed on or near ridges to maximize wind ventilation and should have light-colored reflective roofs.

Pole shelters are much less expensive and more quickly erected than stud-wall structures. They require no foundation and little grading or leveling. Pole construction generally is more desirable for shade because it provides an open, well-ventilated structure.

Stud-wall sheds are more desirable for winter shelter because they have more-tightly-sealed walls and can be insulated. Structures with southern exposure and a dark roof maximize the warmth generated by solar radiation. Stud-wall structures generally last longer and are less susceptible to wind damage. Stud-wall structures may be the only option avail-

able in areas with extremely rocky soils that prevent the digging of post holes for the pole shelter.

Windbreaks

Windbreaks are walls that provide shelter from wind, blowing rain, and sun (Figure 86). Windbreaks are particularly important if livestock are corralled and fed in the winter and are unable to roam to sheltered spots. Obviously, a windbreak should be built on the windward side of an area. A two-sided windbreak that splits the wind at the structure's corner protects livestock effectively.

Windbreaks should be at least eight feet high. Windbreaks can be constructed from one-inch wooden planks, corrugated metal, or concrete panels. Although the latter method generally will last longer, it is more expensive. A local university extension office or conservation district will have specifications for windbreak construction and design.

Providing Water

Providing water on uplands can lure livestock away from riparian areas and is

Figure 86. A two-sided windbreak effectively protects corralled livestock from heavy winds, blowing rain, and sun. (Izaak Walton League of America figure)

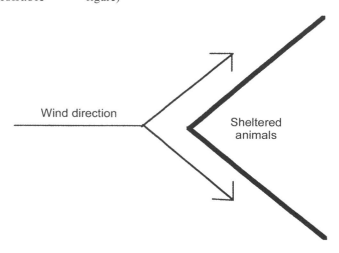

Wind direction

Sheltered animals

particularly important in areas where riparian zones require periods of complete rest. Watering facilities, also called water improvements, are key elements of management plans that seek to emphasize use of upland forage. They facilitate the collection, storage, and distribution of water and make it available to livestock. When combined with intensive upland forage seeding and cultivation, watering facilities can be especially effective at reducing pressure on the riparian area (Figure 87).

Water facilities take three forms: Transport (or trapping) facilities, storage facilities, and drinking facilities (or troughs). Storage facilities are usually not used for drinking by livestock and wildlife, and drinking facilities are not usually big enough to be considered storage facilities. Exceptions exist when a facility contains amounts of water greater than what is needed to meet daily water consumption requirements, but is of a low enough height to allow animals to drink.

Watering facilities require a great deal of deliberation in all aspects of construction and design. The water requirements of the herd, evaporation rates, and other factors must be considered. A cow drinks an average of 15 gallons of water per day. Managers should consider a number of different system options and pipeline configurations to determine the most affordable and effective combination. Remember to consult with the project's technical team when developing watering facilities.

Sources of Water

Wells

Water facilities must be fed by a water source.

Water is piped via the transport system to the storage and drinking facilities. Groundwater wells are one of the most familiar water sources. Wells may not be feasible on uplands because of the costs associated with drilling to the water table. Some groundwater resources may be too sulfurous or too salty to use. A reputable well driller can provide you with information about the efficiency and quality of local wells. Constructing new wells could have an effect on existing wells. Consult with the local water resource agencies before planning a well.

Holding Ponds

Holding ponds are another water source, but they can be expensive to construct and maintain. The ponds collect runoff during the rainy season and from snowmelt. Ponds do not need to be very deep, although deep ponds store more water. In extremely well-drained soils, holding ponds may not be possible. A nontoxic rubber or plastic liner may be necessary to prevent water from infiltrating the soil. Ponds allow silt to settle out of runoff. The cleaner water can be piped to one or more troughs or storage tanks.

Figure 87. This storage tank on Milesnick Ranch provides livestock with an upland water source, keeping them away from sensitive riparian areas. (National Cattlemen's Beef Association photograph)

Runoff collection facilities can be simple or complex, depending on needs and available resources.

Spring Development

Spring development consists of excavating a spring and placing it inside a concrete or wooden enclosure. Concrete boxes can be purchased or constructed with concrete block. Water from the spring builds up inside the box and eventually reaches an outfall that pipes it to the desired water-holding structure. The box (especially wooden boxes) should be covered to protect against **weathering** and fenced off to prevent damage by livestock.

Spring development is a very controversial issue. Natural areas fed by springs should not be deprived of water. The water-holding structure should have an overflow pipe so that water can continue to reach the natural receiving body after filling the holding structure. Many wet upland meadows, which are important sources of forage, can be damaged when deprived of water. Check with state and local water regulatory agencies before considering this approach. It may be illegal depending on your water rights or wetland laws.

Natural Waterways

State regulations also may prohibit the direct piping of water from rivers and streams. This option is not recommended on small streams or streams that do not have a dependable, abundant flow all year. Minimum in-stream flows are necessary to maintain aquatic habitat, water for recreational uses, and water for downstream users.

Drawing water from a stream may be as easy as laying a pipe in the stream and using gravity and the lay of the land to feed water to a holding tank. Once the holding tank has reached the desired capacity, the overflow can be returned to the stream at a location downstream from the intake. Holding tanks should be placed on level ground outside of the riparian zone. One possible design is a piping system used to move water to a trough situated just outside of a fence; this arrangement limits livestock access to the riparian zone. Livestock on uplands can access this water, but cannot enter the riparian zone because of the fence.

Water Transport Facilities

Construction materials and placement of the holding tanks are two important considerations in constructing water transport facilities. The material used for the transport or piping system is very important. Steel pipe is resistant to trampling and other possible animal-related damages, but is susceptible to fracture when temperatures drop below freezing. If metal pipe is desired in places susceptible to freezing, it should be laid as far below the **freeze line** as possible to prevent fracture.

Plastic pipes, including polyvinyl chloride (PVC; white) and polyethylene (PE; black) are somewhat less susceptible to weather damage but are more prone to damage by livestock. If the site allows for underground piping, place the pipe in an area without concentrations of livestock or big game animals, or place it underground at a depth sufficient to protect it from trampling. If possible, plastic pipes should also be buried below the freeze line. Emergent sections of piping systems should be protected from animal damage by a series of posts and stringers that prevent animal contact.

Storage Facilities

Proper site selection is key to the success, efficiency, and longevity of a

water storage facility. Like shelters, storage facilities should not be placed on steep slopes that livestock generally avoid. Try to place storage facilities in areas with well-drained soils. Otherwise, large muddy areas might develop near overflow outlets. It is highly recommended that storage facilities be placed on an erosion-resistant pad to prevent erosion around the facility itself. Gravel is inexpensive and frequently used. Erosion at the base of the tank can cause cracking, tipping, and other maintenance problems in the future.

Sturdier, permanent pads include concrete slabs that can be poured into a wooden form on site. This may not be practical in remote areas. If accumulated manure and soil become sanitation problems, they can be removed easily from the concrete pad.

Cinder-block pads are another more permanent solution. Cinder-block pads should be dug into the ground about two inches and filled with dirt. The blocks should fit tightly against one another. To prevent spread and gaps between blocks, the outside blocks should be pinned in place with rebar. The top of the rebar is crimped over the top of the cinder block to lock the block in place.

Water storage tanks should be located in areas where they can be accessed for maintenance. Algae, dirt, and other debris will accumulate in watering structures. Structures should be thoroughly drained, cleaned, and disinfected at least once a year. In warm, humid areas, mosquito and midge control may be an additional management issue. With proper maintenance and upkeep, watering facilities should provide no health hazards and should not contribute to parasite problems in the herd.

Birds and other animals that attempt to drink from storage tanks may fall in and drown. To avoid this situation and the unsanitary conditions it creates, storage tanks should be designed with escape mechanisms. Rope tied to the side of the tank and anchored to the bottom of the tank with a rock can serve as a ladder. Rough boards placed in the tank at an angle will serve the same function. Floated boards also allow small animals to climb out of the water and rest before they try to escape the tank. Placing a plastic net or screen over the top of the water storage facility may also help.

Drinking Facilities

Drinking facilities, or troughs, are available commercially at a relatively low cost. Many are portable and can be moved easily to accommodate site conditions. Troughs are typically made of galvanized steel and can either be large, cylindrical structures or the more familiar round-bottomed form on legs. Troughs should be large enough to accommodate the daily drinking needs of the resident livestock.

Like storage facilities, troughs should be placed on a concrete or cinder-block pad to protect against erosion. The ground immediately below the outfall should be protected against the force of falling water. To prevent tipping and other animal-related damage, troughs can be placed inside a wooden frame that limits direct contact with animals. Treating the water to prevent parasites, such as bloodworms, is desirable. Troughs should be fitted with wildlife escape devices similar to those used in storage facilities.

One of the easiest and least expensive drinking troughs consists of a section of concrete or steel culvert pipe fitted into a concrete bottom. The bottom is created by placing the pipe end-up on a piece of plastic and filling it with four inches of concrete. The finished product is a durable, long-lasting tank that is transported easily with a tractor or truck. Pouring and setting the bottom of the tank before it is placed into the ground allows the unit to be moved easily and relocated if necessary.

Other Water-Related Considerations

Livestock cannot drink from frozen troughs. In areas where winter access to riparian areas is undesirable, heating devices might be necessary to keep troughs from freezing. Troughs can be painted black to collect solar radiation. Some ranchers have outfitted water troughs with wind vanes that turn small turbines in the water. The stirring action helps prevent freezing.

Water placement is critical in dryer climates and is the principal limiting factor to livestock grazing in many areas of the western United States. A key point to consider is that livestock will not walk more than one mile from water and still use the forage efficiently. Rangeland planners generally discount any forage use in areas located more than one mile from water and agree that livestock will not travel freely more than two or three miles

in search of forage (Figure 88). Piospheres — areas of overgrazing caused by lack of adequate water facilities — develop around watering points, including riparian areas, when distances to the next closest water source are greater than 1.5 miles. Water facility planning calls for adequate water placement, and requires consideration of such factors as topography, availability of water, and costs.

Other Approaches to Providing Water

Most watering facilities are gravity-fed systems that work with the topography of the land to provide water away from naturally occurring water sources, including riparian areas. In some areas, gravity-fed systems may not be feasible. In such cases, mechanically pressurized methods of water transport are necessary. Consult with your local extension office or soil conservation district for information about these options. Various types of multistage and piston pumps are available commercially. Some are driven by wind, gasoline, or electricity. Solar-powered options do not require connection to an electricity source.

Wind-powered pumps are used in many places, but they can be expensive. However, they are long-lived, efficient, and highly cost-effective. Windmills are

Figure 88. These diagrams represent two different approaches to placing water in the pasture and illustrate the resulting effects of forage use for each approach. Shaded areas indicate active forage consumption. Non-shaded areas are too far from water and will be ignored. (Izaak Walton League of America figure)

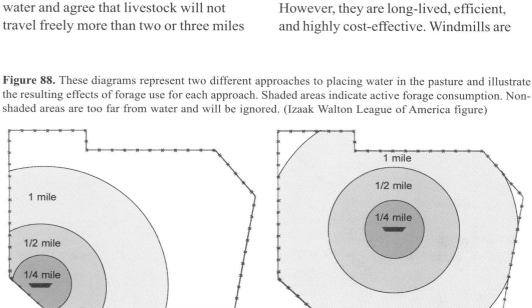

excellent options in remote areas where there is no electricity. Storage facilities can be created adjacent to the windmill to store water. Constructing a windmill requires careful deliberation and planning. The biggest challenge associated with windmills is that they will not work if the wind does not blow.

Water Gaps

In some cases, it is not possible to develop an off-site watering facility in the hopes of providing drinkable water away from riparian areas. In such cases, protected stream access and crossing points are required to reduce damage to the stream bank.

Fence breaks that provide access to the stream for watering and/or crossing are called water gaps. Fencing a stream to restrict livestock access is a common, inexpensive option in areas where some fencing parallels the stream. Modifications are relatively simple and do not require a great deal of time or money.

Water gaps can be designed easily into new fencing systems as well. Water gaps should be combined with some form of barrier to limit livestock migration up and down the stream channel. Corrugated roofing, or sheet metal, suspended from a wire hung between two posts is an effective deterrent. The sheet metal should be about two feet above the surface of the water — low enough to be at an animal's eye level when drinking. If the stream floods, the roofing will swing like a trap door, allowing debris to pass. Water gaps that do not provide for floodwater and flood debris passage may require frequent maintenance and replacement.

Armored water gaps may be particularly useful when livestock are managed in a series of small pastures and must rely on

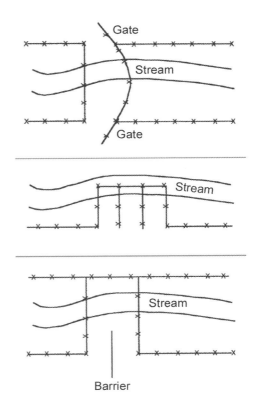

Figure 89. All three of these water gap designs prevent livestock from traveling up and down the stream. The first design is particularly useful when animals must cross the stream to find forage. (Buckhouse, 1995)

free-flowing water. In the East and South, pasture designs often include streams to provide water and a cool retreat from summer heat. Livestock must cross these enclosed creeks often to search for forage, and stream banks can be trampled. While reducing or eliminating crossings is ideal, it may not be possible where pastures are small and waterways are numerous (Figure 89).

A common armoring method used on small streams is a concrete crossing. The crossing is constructed by placing concrete slabs across the stream. Concrete crossings are best suited to shallow headwater streams with gradually sloping banks. The slabs should be installed at an angle less than or equal to that of the channel. Too steep of a gradient will cause eddying and back-cutting, severely damaging the creek. The bank ends of the slabs should be reinforced with riprap to prevent erosion at the crossing points. The riprap reinforcement should extend a minimum of 10 feet beyond the edge of the slab. The crossing should not create a barrier to fish movement up and down the stream. An experienced agricultural engineer should be consulted to investigate this option.

A less expensive and more practical crossing for remote areas consists of wooden logs laid in the stream parallel to the water's flow. Long pieces of rebar are driven into the streambed through pre-drilled holes in the logs. The top of the rebar is crimped over the side of the log to fasten the log in place. Logs should be laid as close together as possible. The structure should extend well beyond the top of the stream banks. If the crossing must be built where stream banks are steep, the portion of the crossing that armors the stream bank can be constructed somewhat like the cribwall discussed in the Bioengineering

Techniques section of Chapter Four. A stair-step design will be necessary. As with concrete crossings, the bank ends should be reinforced with riprap.

Acknowledging Leadership

The Izaak Walton League would like to recognize the following ranches featured in this chapter as leaders of rangeland stewardship.

Dave Wood Ranches, Coalinga, California. Regional Winner of the National Cattleman's Beef Association Environmental Stewardship Award Program.

Milesnick Ranch, Belgrade, Montana. Regional Winner of the National Cattleman's Beef Association Environmental Stewardship Award Program.

Mortenson Ranch, Hayes, South Dakota. Winner of the Environmental Law Institute's 2002 National Wetlands Award.

Conclusion

*H*ealthy streams are critical to providing communities with the economic, ecological, and social benefits that come from clean water. Streams are a valuable indicator of watershed health. They reflect damage from improper land management or natural events. Yet, their dynamic properties make them resilient and create the opportunity for enhancement and recovery.

Cleaning up our country's waters is hard work, so work together. Collaborative approaches lead to better decisions, but building relationships, developing an understanding, and networking are complicated processes that take time. This initial investment is well worth the hard work it requires in order to reduce the potential barriers, red tape, and conflicts that can delay or impede effective and efficient stream stewardship.

The extent of degradation in a watershed may seem daunting, but don't become overwhelmed. It is possible to make a change for the better. This book provides information and ideas to make positive changes for watersheds across the country. Fortunately, the pool of knowledge of stream corridor conditions and water quality indicators is growing rapidly. It is largely up to community groups to put the knowledge to use, initiating enhancement projects and advocating for water resource protection.

Glossary of Terms

Acid Discharge: A discharge with a pH of less than 7.

Active Channel Width: The width of the stream at the bankfull stage.

Aggradation: The geologic process by which streambeds and floodplains are raised in elevation by the deposition of material.

Algae: One-celled or multicellular plants, primarily aquatic.

Aquifer: A body of rock or unconsolidated sediment capable of yielding a usable supply of water.

Bacteria: Free-living unicellular microorganisms of the class *Schizomycetes*.

Bank Erosion Potential Rating: A measurement that indicates the ability of stream banks to resist erosion based on several factors, such as the composition of the bank materials, the bank slope, and the root density of riparian vegetation.

Bank Failure: A slumping off of bank materials when they are not strong or stable enough to resist gravity.

Bank Stability: The likelihood that a stream bank will erode within a defined time frame. While banks erode even in stable river systems, the rate of erosion is much slower in stable streams than in unstable streams. Bank stability is influenced by temperature, composition of the bank material, water flow and force, and the presence of vegetation.

Bankfull: A term that refers to the elevation where, in a stable stream in an alluvial valley, the upper edge of the banks of the stream's channel and the floodplain merge. In streams that are unstable and/or in settings other than alluvial valleys, several alternative indicators of the bankfull elevation occur, some of which are relatively easy to recognize while others require considerable experience or professional assistance to identify. See Bankfull Discharge and Bankfull stage.

Bankfull Discharge: The volume of water, usually expressed as cubic feet per second, that would flow past a given point in its channel if a stream were flowing at bankfull stage. Bankfull discharge — also called bankfull flow, channel-forming discharge, dominant

discharge, and effective discharge — equates with the volume of flow that is most responsible for creating and maintaining a stable stream's channel characteristics. See Bankfull, Bankfull stage, and Discharge.

Bankfull Stage: The elevation of the upper surface of a stable stream in an alluvial valley when the stream has filled its banks and is beginning to spill out onto its floodplain. Bankfull stage is sometimes referred to as bankfull, bankfull elevation, bankfull level, and ordinary high-water mark. See Bankfull, Bankfull Discharge, and Ordinary High-Water Mark.

Base Flow: The portion of stream flow that is derived from natural storage such as springs and groundwater; also considered to be the average stream discharge during low-flow conditions.

Benthic Macroinvertebrates: Bottom-dwelling or substrate-oriented organisms that are large enough to see without magnification and do not have a backbone.

Bioengineering: An applied science that uses structural materials and living plants to control erosion, sedimentation, and flooding.

Biological Monitoring: The use of benthic macroinvertebrates, plants, amphibians, fish, or other living organisms to assess or monitor environmental conditions.

Brace Bar: A bar that runs between the brace posts and corner posts of a cattle fence to help stabilize the corner of the fence. Two brace posts with a brace bar forming an "H" also are used at regular intervals along the length of the fence, usually every quarter of a mile and near gates, to provide strength and prevent sag.

Brace Post: One of the four (sometimes more) posts placed (two on each side) adjacent to and supporting each corner post of a fence that encloses a cattle pasture.

Branch Packing: A bioengineering technique for stabilizing gullies that involves layering dead branches and mulch in the gully of a small eroding section of a bank and fastening them down with live stakes and wire.

Brush Layer: Live branch cuttings crisscrossed on trenches between successive benches of soil.

Buffer: A vegetated area of grass, shrubs, or trees designed to capture and filter runoff from surrounding land uses.

Channel: A natural or artificial waterway of perceptible extent that periodically or continuously contains moving water. It has both a bed and banks that serve to confine the water.

Channel Cross-Section: A graph or plot of ground elevation across a stream valley or a portion of it, usually showing a slice, or cross-section, of the stream channel along a plane oriented perpendicular to the direction of flow.

Channel Roughness: Irregularities in the physical elements of a stream channel upon which energy is expended, including coarseness and texture of bed material, the curvature of the channel, and variation in the longitudinal profile.

Channel Slope: See Slope.

Chronic Toxicity: A condition that causes harm to a life form over a prolonged exposure to toxic substances and has cumulative effects on growth, respiration, or reproduction.

Chute: A new channel formed across the base of a meander.

Coir: A coconut-fiber mesh that can be used to hold exposed soil in place until new plants are established.

Contour Line: A line on a map that joins points of equal elevation.

Corner Post: A post that supports a cattle fence at the corners. Corner posts must be strong, sturdy, and well anchored.

Cribwall: A hollow structure used for bank and slope stabilization formed by mutually perpendicular and interlocking members (usually timber) into which live cuttings and soil are inserted to stabilize roots.

Cross-Section: See Channel Cross-Section.

Culvert: A sewer or drain crossing under a road or embankment.

Deferred Rotation System: A pasture management practice that allows grazing to occur after vegetation has reached a desired stage of development. For example, it may be desirable to delay grazing until after grasses fully develop seed heads.

Deflector: A structure used to deflect stream flow to a different location, usually away from an eroding bank.

Deposition: The natural process by which stream water lays down a deposit of sediment. Also called sedimentation.

Dimension: The cross-sectional area of a stream (width multiplied by average depth).

Discharge: The volume of water flowing past a given point over a defined period of time. In the United States, discharge is usually expressed as the volume of water, measured in cubic feet, that moves past a given point each second — that is, as cubic feet per second (cfs). Discharge also is referred to as stream flow.

Drainage Basin: See watershed.

Dynamic Equilibrium: In reference to stream characteristics, the balance between the amount of water and sediment leaving a stream segment and the amount of water and sediment entering the stream segment.

Effluent: Water or any other fluid that flows out of someplace, such as from groundwater into a body of surface water or from one body of surface water into another.

Enhancement: In the context of restoration ecology, any improvement of a structural or functional attribute; putting back into good condition or working order.

Entrenched Stream: A stream whose channel has been eroded to a depth that exceeds the bankfull stage of the stream. See Incised Stream.

Entrenchment Ratio: The degree of vertical containment of a stream channel. The measurement is the ratio of the flood prone width (double the height of the bankfull stage) by the bankfull width.

Ephemeral Stream: A stream that flows only for very short periods during and immediately after rainfall.

Erosion: The wearing away of soil and other materials through natural and unnatural causes.

Evaporation: The process of a liquid, such as water, being transformed into a gas.

Exotic Species: A species of microbe, plant, or animal that is not native (i.e., is alien) to a particular ecosystem. The introduction of an exotic species might cause ecological and/or economic harm because exotic species sometimes have no natural predators in the new ecosystem to keep them in check.

Fascine: A sausage-like bundle of plant cuttings used to stabilize stream banks and other slopes. Also known as a wattling.

Fill: Soil or other material placed as part of a construction activity.

Floodplain: The flat area of land adjacent to a stream that is covered by the stream during floods.

Flow: The movement of surface water or groundwater in response to gravity or pressure.

Food Chain: An arrangement of the organisms of an ecological community according to the order of predation in which each uses the next usually lower member as a food source.

Forb: Any broad-leaved herbaceous plant other than those in the Gramineae (Poaceae), Cyperaceae, and Juncaceae families.

Freeze Line: The level below which the ground does not freeze.

Fry: Immature fish that still have a yolk sac.

Gabion: A wire basket filled with rocks used to stabilize stream banks and to control erosion.

Gradient: Slope calculated as the amount of vertical rise over horizontal run.

Groundwater: Water that is stored in pores, cracks, and crevices below the water

table and that serves as the source water for wells and springs.

Habitat: The area or environment in which a population or individual lives. The habitat includes not only the place where an individual, population, or species is found, but also the characteristics of that place that meet the life cycle needs of that species.

Headwaters: The uppermost reaches of a stream or river.

Herbaceous: Plants with non-woody stems.

Hydraulics: Water or other liquids, static or in motion, and their actions. The study of practical applications of liquid in motion. The science and technology of the static and dynamic behavior of fluids.

Hydric Soil: A soil found in saturated, anaerobic environments and usually characterized by a gray or mottled appearance; often found in wetlands.

Hydrology: The study of the properties, distribution, and effects of water on Earth's surface, soil, and atmosphere.

Hydrophyte: A plant adapted to living in wet conditions.

Imbricated Riprap: Stones or crushed rock that have been placed along a stream bank to stabilize the bank. The stones of imbricated riprap are contiguous to each other, which requires that they be put in place by hand, rather than dumped onto the bank as loose stones.

Impervious: Not allowing passage or movement of a fluid into or through. Synonymous with impermeable.

Incised Stream: A stream that has cut into the streambed, lowering its channel floor below the level of its former floodplain. See Entrenched Stream.

Influent Stream: A stream that loses water to groundwater.

Intermittent Stream: A stream that has interrupted water flow or does not flow continuously.

Invasive Species: A species of microbe, plant, or animal that is not native (i.e., is exotic, alien, or non-native) to a given ecosystem but whose presence in that ecosystem causes or might cause economic or environmental harm to the system or harm to human health. See Exotic Species.

Island Bar: An island that forms in the middle of a stream where water flows too slowly to transport the available sediment.

Jute Mesh: Netting made of plant fiber similar to burlap that can be used to hold exposed soil in place until new plants are established.

Levee: An embankment, berm, or low ridge of sediment alongside a waterway resulting from near-stream deposition during floods. Natural levees form when sediment-laden water spills over the banks of streams that flood. The sudden loss of depth and velocity causes sediment to drop out of suspension and collect along the edge of the stream. Artificial levees are constructed to restrict streams or other waterways from overflowing their banks.

Line Post: A post used in cattle fencing, line posts are placed at regular intervals along the wire to hold the wire up, maintain wire spacing, and prevent sag and tension loss.

Live Stake: A cutting from a live branch that is inserted into the soil to stabilize the slope of a stream bank when the cutting roots and grows. Dead stakes, also wooden, can also be used to keep other stream bank stabilization aids in place.

Longitudinal Profile: A side-view representation of the stream along its length.

Macroinvertebrate: A spineless animal visible to the naked eye or larger than 0.5 millimeter. Benthic macroinvertebrates live in the bottom of streams and wetlands.

Meander: A circuitous winding or bend in the river.

Meander Scroll: A natural sedimentary feature that forms as meandering stream channels migrate laterally across the floodplain. Meander scrolls indicate the former location of stream channels that have mostly filled in with sediment.

Morphology: The shape of things, such as channels, stream corridors, watersheds, and other topographic features.

Nick Point: The upstream-most point where the stream is actively eroding to a new base level. Nick points migrate upstream.

Non-Point Source Pollution: Pollution that originates from diffuse sources and usually is not regulated. Runoff from streets that contains oil, feces, and other sediments is an example of non-point source pollution.

Non-Structural: With respect to enhancing stream conditions, refers to modifications of circumstances, such as land use, that affect the runoff or infiltration of water that eventually enters the stream. A grazing management plan or the addition of a stream-side forest buffer are examples of non-structural stream enhancement techniques.

Ordinary High-Water Mark: A term often used by federal and state agencies to describe the water level reached during bankfull stage. See Bankfull Stage.

Oxbow: A meander that was severed from the main channel when a new channel formed across the base of the meander.

Oxbow Lake: A former meander in the river, still containing water, that is cut off from the main channel by deposited sediment.

Pattern: Refers to the "plan view" of a channel as seen from above.

Perennial Stream: A stream that flows continuously.

Point Bar: A gravel or sand deposit on the inside of a bend in a river.

Point Source Pollution: Pollution that is discharged from an identifiable point of origin, sometimes through a pipe or other conduit, and is usually regulated by either the state or federal government.

Pollutant: Any substance that causes harm to human health or to the environment.

Pool: Deeper areas of a stream with slow-moving water that are often used by larger fish for cover.

Reach: A section of stream (defined in a variety of ways, such as the section between tributaries or a section with consistent characteristics).

Reference Reach: What a degraded stream site could have looked like had it remained stable. A reference reach shares certain characteristics with the project site, including similar size, stream type, location in the landscape, and surrounding land uses. A reference reach is not always pristine, but it should represent what is reasonably attainable.

Rest Rotation System: A pasture management technique that allows each pasture in the system at least one year of complete rest. Livestock forage in other pastures or receive supplemental feed. Rest rotation normally involves a minimum of three pastures.

Restoration: The return of an ecosystem to a close approximation of its condition prior to disturbance.

Revetments: Facing of stone or other material placed along the edge of a body of water to stabilize the bank and protect it from erosion.

Riffle: A shallow section in a stream where water is breaking over rocks or other partially submerged organic debris and producing surface agitation.

Riparian Zone: The land area adjacent to a stream or any other water body.

Riprap: Stones of varying size used to stabilize stream banks and other slopes.

River: A large, natural stream of water that runs into a larger water body such as a lake or the ocean.

Runoff: The portion of precipitation on land that flows over the ground surface. It can collect nutrients, pollutants, and other dissolved or suspended materials from air or land and carry them to the receiving waters.

Scarp: An escarpment, usually a steep slope or cliff caused by erosion. This term is also used to describe the stream bank.

Scour: The erosive removal of material from the streambed and banks.

Sediment: Fragmentary material that originates from the weathering of rocks. Vast amounts of sediment are transported and deposited by running water.

Sinuosity: The amount or degree of curvature of a stream channel.

Site Inventory: An in-depth documentation of the conditions at a particular segment of a stream. Data collected may include bank stability, channel condition, riparian vegetation, canopy cover, instream fish cover, barriers to fish movement, and other parameters.

Slope: The amount of vertical rise divided by horizontal run.

Slumping: The collapse of slopes caused by undercutting.

Spherical Densiometer: A piece of equipment used to estimate the percentage of tree canopy cover. It consists of a concave mirror with a grid of equally spaced dots or lines on the surface. The mirror reflects the vegetation overhead and percentage of canopy cover is calculated by dividing the number of dots covered by the reflection of vegetation by the total number of dots.

Splay: A delta-shaped deposit of coarser sediments that occur next to a stream when a natural levee is breached. Natural levees and splays can prevent floodwaters from returning to the channel when floodwaters recede.

Step: Part of the typical stair-step structure of steep mountain streambeds, commonly referred to as steps and pools. Steps are formed from resistant bedrock, cobbles, and boulders that lie across the channel. The steps are separated by deep pool areas.

Storm Drain Outfall: The place where a storm drain discharges. Often, the outfall takes the form of a pipe or other structure through which water collected from streets and buildings is discharged into a stream or other water body.

Storm Flow: Precipitation that rolls over the land (runoff) or through the ground (throughflow) to the channel over a short time frame and can cause flooding.

Stream: A generalized term for a natural body of running water moving over Earth's surface in a channel or bed. Rivers, creeks, brooks, and runs are all streams.

Stream Bank: The portion of the channel cross-section that restricts lateral movement of water at normal water levels.

Stream Bank Stabilization: The process of supporting the structural integrity of earthen stream-channel banks to prevent bank slumping, undercutting of riparian trees, and overall erosion.

Stream Buffer: A variable-width strip of vegetated land adjacent to a stream that is preserved from development activity to protect water quality and aquatic and terrestrial habitats.

Stream Channel Degradation: Geologic process by which a stream bottom is lowered in elevation due to the net loss of substrate material. Also called downcutting.

Stream Corridor: The area that includes the stream channel, floodplain, and transitional upland fringe.

Stream Flow: See Discharge.

Stream Restoration: Various techniques used to replicate the hydrological, morphological, and ecological features that have been lost in a stream due to urbanization, agriculture, or other disturbance.

Structural: With respect to enhancing stream conditions, refers to physical modification of the stream bank and often includes adding materials such as rock to harden the bank.

Substrate: The mineral or organic material that forms the bed of the stream.

Succession: The unidirectional change in the composition of an ecosystem as the available competing organisms, especially plants, respond to and modify the environment.

Terrace: A bench-like landform with a nearly level plain bounded by rising and falling slopes. Streams often have one or more terraces on either side of the floodplain. In cross-section, terraces resemble stairs moving up and away from the stream channel or floodplain.

Thalweg: The line connecting the lowest or deepest points along the longitudinal axis of the streambed.

Toe: The bottom of a slope or bank.

Toxic: Something that is harmful, destructive, or deadly.

Transpiration: The discharge of water, in a gaseous state, through the skin of animals or the pores of plants.

Turbidity: Murkiness or cloudiness of water caused by suspended particles such as fine sediments (silts, clays) and algae.

Vegetative: Of or pertaining to the plant community and its inherent characteristics, such as diversity, structure, amount of ground cover provided, and ecological functions. With respect to enhancing stream conditions, refers to planting vegetation on or adjacent to the stream banks.

Velocity: The rate of motion of objects or particles, or of a stream of particles. Velocity is usually represented as change in distance over time.

Watershed: The total area of land from which water drains into a stream or other body of water.

Watershed Assessment: Evaluates the condition of a watershed including its chemical, biological, and physical features. This assessment can be used to identify enhancement priorities, reference reaches, and potential effects of proposed development on stream-channel stability.

Wattling: See Fascine.

Weathering: The act of cracking, crumbling, flaking, or dissolving of materials caused by the action of wind, air, and water.

Wetland: An area of land that is saturated at least part of the year by water. Usually found in depressions or low-lying areas on floodplains or coastal areas.

Width-to-Depth Ratio: A measurement of the average bankfull width of a stream over the average bankfull depth of the stream. This ratio can provide clues about potential future erosion.

Zoning: The practice of dividing land into parcels according to the uses or activities that occur or are allowed within it.

How to Read a Topographic Map and Delineate a Watershed

*I*n order to successfully delineate a watershed boundary, the evaluator needs to visualize the landscape as represented by a topographic map. This is not difficult once the following basic concepts of topographic maps are understood.

Each contour line on a topographic map represents a ground elevation or vertical distance above a reference level, such as sea level. A contour line is level with respect to Earth's surface just like the top of a building foundation. All points along any one contour line are at the same elevation.

The difference in elevation between two adjacent contours is called the contour interval. This is typically given in the map legend. It represents the vertical distance you would need to climb or descend to get from one contour elevation to the next.

The horizontal distance between contours, on the other hand, is determined by the steepness of the landscape and can vary greatly on a given map. On relatively flat ground, two 20-foot contours can be far apart. On a steep cliff face two 20-foot contours might be close together. In each case the vertical distance between the contour lines would still be 20 feet.

One of the easiest landscapes to visualize on a topographic map is an isolated hill. If this hill is more or less circular, the map will show it as a series of more or less concentric circles (Figure A-1). Imagine that a surveyor actually marks these contour lines on the ground. If two people start walking in opposite directions on the same contour line, beginning at point A, they will eventually meet face to face.

If these same two people start out in opposite directions on different contours, beginning at points A and B respectively, they will pass each other somewhere on the hill and their vertical distance apart would remain 20 feet. Their horizontal distance apart could be great or small depending on the steepness of the hillside where they pass.

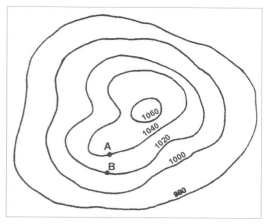

Figure A-1. Isolated hills are depicted on a topographic map as a series of concentric circles.

A rather more complicated situation is one where two hills are connected by a saddle (Figure A-2). Here each hill is circled by contours but at some point toward the base of the hills, contours begin to circle both hills.

How do contours relate to water flow? A general rule of thumb is that water flow is perpendicular to contour lines. In the case of the isolated hill, water flows

Figure A-2. When a saddle connects two adjacent hills, contour lines near their bases encircle both hills.

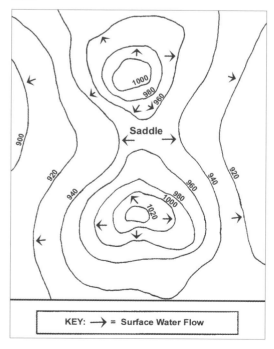

KEY: → = Surface Water Flow

down on all sides of the hill. Water flows from the top of the saddle or ridge, down each side in the same way water flows down each side of a garden wall.

As the water continues downhill it flows into progressively larger watercourses and ultimately into the ocean. Any point on a watercourse can be used to define a watershed. That is, the entire drainage area of a major river like the Merrimack can be considered a watershed, but the drainage areas of each of its tributaries are also watersheds.

Each tributary in turn has tributaries, and each of these tributaries has a watershed. This process of subdivision can continue until very small, local watersheds are defined which might only drain a few acres, and might not contain a defined watercourse.

Figure A-3 shows the watershed of a small stream. Water always flows downhill perpendicular to the contour lines. As one proceeds upstream, successively higher and higher contour lines first parallel then cross the stream. This is because the floor of a river valley rises as you go upstream. Likewise, the valley slopes upward on each side of the stream. A general rule of thumb is that topographic lines always point upstream. With that in mind, it is not difficult to make out drainage patterns and the direction of flow on the landscape even when there is no stream depicted on the map. In Figure A-3, for example, the direction of streamflow is from point A to point B.

Ultimately, you must reach the highest point upstream. This is the head of the watershed, beyond which the land slopes away into another watershed. At each point on the stream the land slopes up on each side to some high point then down into another watershed. If you were to join

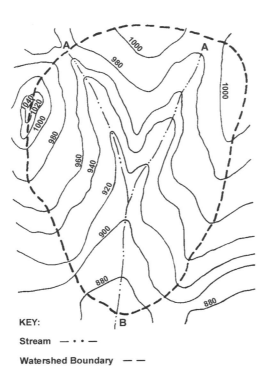

KEY:

Stream   — · · —

Watershed Boundary   — —

Figure A-3. When identifying drainage patterns of a stream system on a topographic map, remember that contour lines always point upstream. In the watershed depicted, water flows from points A to B.

all of these high points around the stream you would have the watershed boundary. (High points are generally hill tops, ridge lines, or saddles.)

Delineating a Watershed

The following procedure and example will help you locate and connect all of the high points around a watershed on a topographic map shown in Figure A-4. Visualizing the landscape represented by the topographic map will make the process much easier than simply trying to follow a method by rote.

1. Draw a circle at the outlet or downstream point of the wetland in question (the wetland is the hatched area shown in Figure A-4).

2. Put small "Xs" at the high points along both sides of the watercourse, working your way upstream towards the headwaters of the watershed.

3. Starting at the circle that was made in step 1, draw a line connecting the "Xs" along one side of the watercourse (Figure A-5). This line should always cross the contours at right angles (i.e., it should be perpendicular to each contour line it crosses).

4. Continue the line until it passes around the head of the watershed and down the

Figure A-4. The first step in delineating a watershed boundary is to identify and mark the outlet point of the stream, and all of the high points around it.

Figure A-5. The watershed boundary marks the outer limits of a land area that drains into a single stream system.

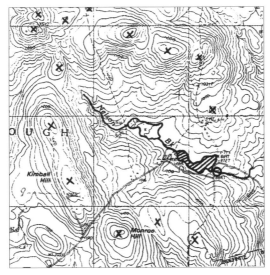

opposite side of the watercourse. Eventually it will connect with the circle from which you started. At this point you have delineated the watershed of the wetland being evaluated.

The delineation appears as a solid line around the watercourse. Generally, surface water runoff from rain falling anywhere in this area flows into and out of the wetland being evaluated. This means that the wetland has the potential to modify and attenuate sediment and nutrient loads from this watershed as well as to store runoff which might otherwise result in downstream flooding.

Measuring Watershed Areas

There are two widely available methods for measuring the area of a watershed: The Dot Grid Method and the Planimeter.

1. The dot grid method is a simple technique which does not require any expensive equipment. In this method the user places a sheet of acetate or mylar, which has a series of dots about the size of the period at the end of this sentence printed on it, over the map area to be measured. The user counts the dots which fall within the area to be measured and multiplies these by a conversion factor to determine the area. A hand-held, mechanical counting device is available to speed up this procedure.

2. The second of these methods involves using a planimeter, which is a small device having a hinged mechanical arm. One end of the arm is fixed to a weighted base while the other end has an attached magnifying lens with a cross hair or other pointer. The user spreads the map with the delineated area on a flat surface. After placing the base of the planimeter in a convenient location the user traces around the area to be measured with the pointer. A dial or other readout registers the area being measured.

Planimeters can be costly depending on the degree of their sophistication. Dot counting grids are significantly more affordable. Both planimeters and dot grids are available from engineering and forestry supply companies. Users of either of these methods should refer to the instructions packaged with the equipment they purchase.

[This appendix is from Appendix E of the *Method for the Comparative Evaluation of Nontidal Wetlands in New Hampshire* (Ammann and Stone, 1991). This document and method are commonly called "The New Hampshire Method." It is reprinted here, with some modifications, from the Natural Resources Conservation Service state office in Durham, New Hampshire.]

Establishing a Technical Team for Stream Enhancement Projects

*A*n interdisciplinary team of specialists provides the knowledge, skill, ability, and professional judgment a project needs to succeed. Experts often have access to information needed for assessment, site selection, inventory, engineered design, permits, installation, monitoring, and maintenance. Experts need to be selected before the project gets off the ground, so begin finding these people early. Building relationships takes time. The experts you find — whether in government agencies, schools and universities, consulting firms, or other organizations — vary depending upon who is locally available and the amount of funding obtained.

The technical advisory team should include people who can help solicit financial support and coordinate public outreach. Government agency officials can assist with the acquisition of necessary permits. Professional advisors might be able to help gather information needed to

identify problems and appropriate solutions, including site analysis, design and implementation of stabilization measures, and monitoring. Experts can also investigate sensitive legal, economic, or cultural issues that might influence the enhancement effort.

Which Experts Should Be Part of the Technical Team?

Hydrologists: Hydrologists study precipitation and the flow of water on the Earth's surface and underground. Specialized hydrologists called "fluvial geomorphologists" can help you understand stream flow conditions and how a natural stream channel can change over time. They examine the erosion cycle of land depositions and degradation in rivers. You can find a hydrologist or fluvial geomorphologist through local offices of the Geological Survey, Army Corps of Engineers, Natural Resources Conservation Service, state fish

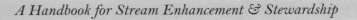

and game agencies, ecological restoration consulting firms, or universities.

Landscape architects and civil engineers: Many civil engineers understand stream mechanics and dynamic ecological systems. They can offer knowledge on constructing biological systems to stabilize stream corridors. Landscape architects who also have experience in bioengineering can help design your project to incorporate a multi-use strategy, such as providing for fish and wildlife habitat, nature trails, or recreation. These specialists work with consulting firms, government agencies, and local conservation organizations.

Botanists: A botanist identifies plants and characterizes the major plant communities in the riparian buffer along the stream. Volunteers can learn to identify many riparian plants by practicing with field guides. Botanists also can locate and protect rare and endangered plants and make recommendations about native plants for stabilizing banks, improving canopies, and providing food for wildlife. They can usually direct you to sources of native vegetation. To find a botanist, contact native plant societies, garden clubs, local consulting firms, universities, or state or local environmental agencies. A landscape architect who is familiar with riparian vegetation also can help you identify plants and assist with sketching and characterizing the site.

Biologists, zoologists, aquatic ecologists, or ***entomologists:*** Biologists identify animals living in or using streams and the riparian zone, and can explain the ecological needs of all native species. Biologists can be found through consulting firms, universities, or state or local environmental agencies. A local bird club or Audubon Society member can help identify riparian birds. An entomologist from a local college, university, or conservation association can help to identify macroinvertebrates that live in the stream. Professionals with similar skills are also on the staff of many agencies.

Soil scientists: Soil scientists understand the nature of the soils along the stream and can evaluate the potential for future erosion and bank failure. They can also help determine if the floodplain and banks might consist of unstable fill material or have contamination problems.

Computer experts: Computer specialists might be able to help establish a database that models stream problems and various enhancement options. They can calculate the floodwater retention capacity of different plans. They can also help you pinpoint areas in the watershed where strategically placed stormwater management ponds or constructed wetlands can reduce or slow stormwater flows.

Chemist: A professional chemist or staff member of a laboratory can help with the collection, review, and analysis of water quality samples and data. Local universities, state agencies, commercial laboratories, county public health departments, or bottled water companies could volunteer lab services. Contact your state water regulatory agency to inquire about laboratory facilities and services.

Other professionals that could provide important input for the interdisciplinary effort include foresters, legal consultants, engineers, economists, archaeologists, sociologists, rangeland specialists, public involvement specialists, real estate experts, and Native American tribal leaders.

Where Do You Find These Experts?

Experts from local environmental organizations, government agencies, consulting firms, or universities — people with knowledge and experience doing enhancement projects — are often willing to help with community projects, perhaps as volunteers.

Federal agencies might have offices in your area that can offer information, funding, or technical assistance for stream enhancement projects. Agencies involved in stream conservation include the Fish and Wildlife Service, the Bureau of Land Management, the National Park Service, the National Marine Fisheries Service, the National Biological Service, the Environmental Protection Agency, the Army Corps of Engineers, the Tennessee Valley Authority, the Geological Survey, the Bureau of Reclamation, the Natural Resources Conservation Service, the Forest Service, and the Cooperative Extension Service.

The Natural Resources Conservation Service provides a list of local service centers on its web site (*www.nrcs.usda.gov*). It also provides information on weather, water management, watershed planning, and flood control. The Resource Conservation and Development program focuses on improvement of quality of life achieved through natural resources conservation and community development. Natural Resources Conservation Service can provide grants for land conservation, water management, community development, and environmental needs in designated Resource Conservation and Development areas.

The Fish and Wildlife Service's Partners for Fish and Wildlife program works with state and local agencies to provide financial and technical assistance to private landowners to enhance private lands. If a project is adjacent to or near public lands, the public land manager can be your most valuable government partner by providing technical or financial assistance. The National Park Service's Rivers Trails and Conservation Assistance Program can help organize stakeholder meetings, identify enhancement and recreation opportunities, and create a plan for implementing the project and securing funding.

Fifteen federal agencies and other watershed groups created the *Stream Corridor Restoration* handbook (*www.usda.gov/stream_restoration*) to give technical help to groups restoring stream corridors. Most county and local governments have experts who can help design effective projects. Departments of parks and recreation, soil and water conservation districts, and agricultural extension offices all could be of assistance locally.

Contact local universities that offer courses in stream ecology, hydrology, geology, botany, or other areas of expertise pertinent to your project. Professors and graduate students may be willing to provide advice.

Professional consultants also are a great resource for a stream enhancement technical team. Restoration ecologists may have a wide range of experience that would benefit your enhancement project. While these consultants generally are paid for this work, some may be willing to help as volunteers or give your group a discount.

In addition to professional consultants and government staff, expertise is often available through local nonprofit organizations that have done enhancement projects (e.g., chapters of the Izaak Walton League, National Audubon Society, Trout Unlimited, or National Wildlife Federation). Members of these groups could have expertise in stream ecology or may know others who can help. If funding is available, consider hiring a consultant to help assess the stream and develop possible solutions.

The Izaak Walton League provides technical assistance to community groups across the country through its Save Our Streams web site (*www.iwla.org/sos*) and toll-free hotline (800) BUG-IWLA. Other good places to look are The Conservation Directory, published by the National Wildlife Federation; the *National Directory of Volunteer Environmental Monitoring Programs*, available on-line from the Environmental Protection Agency (*yosemite.epa.gov/water/volmon.nsf*); and the Environmental Protection Agency's *Surf Your Watershed* web page (*www.epa.gov/surf*). The Watershed Information Network (*www.epa.gov/win*) is an on-line resource with links to watershed partners, including federal and state agencies and local watershed groups. The Environmental Protection Agency's Adopt Your Watershed program (*www.epa.gov/adopt*) and River Network (*www.rivernetwork.org*) both provide assistance to watershed groups.

The Environmental Support Center (*www.envsc.org*) helps grassroots groups improve their management, planning, funding, organizing, and communications capabilities through trainings, technological resources, subsidies, and low-interest loans. The Environmental Support Center also coordinates a national network of 50

to 60 state environmental councils, which are improving environmental advocacy at the state level. The nonprofit Community and Environmental Defense Services (*www.ceds.org*) is another helpful resource that teaches people how to organize their community to stop an environmental threat, including poorly planned housing projects, shopping centers, highways, landfills, mining activities, and so on. Community and Environmental Defense Services is a combination of a legal clinic and a consulting group that consists of a nationwide network of attorneys, environmental scientists, traffic engineers, land-use planners, political strategists, fundraisers, and other professionals. For more resources, visit the League's Watershed Stewardship Resources document available on-line at *www.iwla.org/sos/resources*.

It is important to note that not all professionals will be able to volunteer their time to help citizen groups. Before asking agency staff to help with a stream enhancement, spend time learning about the stream project. Use a technical team to answer specific questions like, "Could you please help us identify exotic species of plants that are present in our region?" instead of "What are exotic species?" Some agency biologists have to schedule their time a year in advance, so their availability for projects may be limited. Keep expectations and requests reasonable.

Make sure to follow up with thank-you calls or letters and take time to make a personal visit to strengthen relationships with technical team members and others who have helped your group. It may open doors to other sources of useful information.

Appendix C

Stream Channel Stability Assessment

*A*n assessment of existing conditions is the vital first step in planning a stream enhancement project. A watershed assessment provides the information needed to choose appropriate restoration sites. A site inventory and assessment provides information needed to choose enhancement techniques.

In this Appendix, we provide a site inventory and channel stability assessment based on a method described in *Stream Reach Inventory and Channel Stability Evaluation (*Pfankuch, 1975). This assessment method looks at the stability of upper and lower stream banks; the size and composition of materials on the stream bottom; the width, depth, and velocity of the stream; and other factors to determine conditions that exist at a specific stream site. Sites are scored on a variety of indicators and are given a score of excellent, good, fair, or poor. Sites that score in the excellent or good range may require protection to help sustain their current

channel stability. Sites that score fair or poor may be candidates for enhancement or restoration. Individual scores for each channel stability indicator assessed provide further information on which restoration and enhancement techniques may be most appropriate. This assessment also may be used to prioritize sites for restoration by determining stream channel stability ratings for multiple sites within the same watershed.

Volunteers may need technical help to perform this assessment because it does require some ability to recognize stream corridor characteristics, such as bankfull stage. It is suggested that volunteers avoid keying in on a single indicator or group of indicators when rating stream channel stability. The total rating score made by an inexperienced person is often numerically close to the score a more experienced person would give a particular stream. When an inexperienced person over-rates and under-rates specific parameters, the total score tends to balance out.

The assessment form and instructions are given below. For more examples of watershed assessments and site inventories and assessments, please see the Izaak Walton League's Watershed Stewardship Resources Listing at *www.iwla.org/sos/resources.*

Stream Channel Stability Assessment Data Form and Instructions

SITE INFORMATION

Stream Name: _____

Date: _____   Time: _____

Site Location and Description: _____

_____

_____

This site assessment provides a score for each of fifteen separate indicators of bank stability. After evaluating each indicator and circling the associated scores, the user adds all the scores to determine the overall stability of the stream channel as either excellent, good, fair or poor. First, look at the Key to Stability Indicators for a detailed description of each indicator and where to look for the indicator in the stream corridor (Table A-1). Next, complete the Stream Channel Stability Scoring Sheet (Table A-2). The first column provides a stability indicator number that matches the numbers on the Key to Stability Indicators. For each stability indicator, examine the descriptions listed on the Stream Channel Stability Scoring Sheet under the columns titled excellent, good, fair and poor. Determine which description best fits the condition at the stream site and circle the corresponding number of points for each stability indicator. For more information on terms used, see the Glossary. At the end of the sheet, total up the scores for each column. Transfer the total scores for the excellent, good, fair, and poor columns below, and add them all to determine a total score for the site. Overall stream channel stability ratings are provided below.

Excellent _____ + Good _____ + Fair _____ + Poor _____

=   Total Site Score _____

Stream Channel Stability Ratings:*   <38   =   Excellent
76   =   Good
77–114   =   Fair
115 + =   Poor

*Hydrology professionals may adjust the scores above.

Table A-1. Key to Stability Indicators.[1]

Stability Indicator	Description	Where To Look
1	Bank slope: All other factors being equal, the steeper the slope, the greater the potential for erosion.	Upper bank[a]
2	Evidence of slope failure by mass wasting: Mass wasting is mass movement of banks by slumping or sliding. Mass wasting introduces large volumes of soil into the channel suddenly.	Upper bank
3	Woody debris in the channel: Excessive amounts of woody debris in the channel may form a debris jam that causes bank erosion as water tries to flow past the obstruction.	Upper bank
4	Density of vegetation: Estimate depth and density of root mass based on presence of vegetation on or near the streambanks. Dense vegetation stabilizes banks.	Upper bank
5	Channel capacity to hold existing stream flows and storm flows: Also looks at width-to-depth ratio, which may indicate the channel is widening or deepening.	Lower bank[b]
6	Bank rock content: The volume, size and shape of the rocks within the stream bank reveal how well the bank will resist erosion.	Lower bank
7	Debris jam: Rocks and logs cause flow to be diverted and erode the banks. Worst case is that the obstructions can move down the stream and induce multiple erosion points. Best scenario is when the obstruction is firmly embedded and does not significantly alter stream flow.	Lower bank
8	Bank erosion: Evidence and height of raw banks.	Lower bank
9	Point bar enlargement: Evidence of enlargement can be absence of vegetation and an abundance of sand.	Lower bank
10	Substrate shape: Smooth rocks indicate that the channel bed material is unstable, sharp edges have been rounded off as the material has been moved around. Sharp angular rocks indicate a more stable channel.	Channel bottom[c]
11	Substrate staining: Rocks that are stained indicate that they have been in place for a long period of time, hence the channel is stable. Bright rocks have been polished by frequent tumbling and indicate less stability.	Channel bottom
12	Substrate imbrications: Tightly packed substrate indicates a stable bed. Under stable conditions, voids are filled by smaller rocks and particles. Larger components tend to overlap like shingles (imbricate) and are very stable.	Channel bottom
13	Substrate size distribution and the percent of substrate that is moveable: Visual estimation of the variation in sediment size. This includes an assessment of changes or shifts from the natural variation of particle sizes for a given stream type and the percentage of all components that are judged to be stable	Channel bottom

(continued on next page)

Table A-1 (continued)

Stability Indicator	Description	Where To Look
	materials. Bedrock, large boulders, and stones of one to three feet in diameter are generally considered stable. Smaller rocks may also be considered stable in the smaller channels. Bedrock always results in an excellent classification.	
14	Bed scouring and deposition: Looking for evidence of scour holes and sand bars. Estimate the percentage of bed that has been scoured or deposited.	Channel bottom
15	Algae: Both brown and green algae indicate stability because rocks that stay in place on the channel bottom allow algae to grow.	Channel bottom

[1] This key provides information needed to complete the Stream Channel Stability Scoring Sheet. Each stability indicator is given a number that also is used on the scoring sheet. This table provides a detailed description of each indicator and information about where to look for the indicator within the stream channel.

[a] Upper bank — From the break in the general slope of the surounding land to the normal high water line. Terrestrial plants and animals normally inhabit this area.

[b] Lower bank — The intermittently submerged portion of the channel cross section from the normal high water line to the water's edge during low flow periods.

[c] Channel bottom — The submerged portion of the channel.

Table A-2. Stream Channel Stability Scoring Sheet[1]

Stability Indicator Numbers	Rating							
	Excellent		Good		Fair		Poor	
1	Bank slope gradient <30%.	2	Bank slope gradient 30–40%.	4	Bank slope gradient 40–60%.	6	Bank slope gradient >60%.	8
2	No evidence of past or any potential for future mass wasting into channel.	3	Infrequent and/or very small. Mostly healed over. Low future potential.	6	Moderate frequency and size with some raw spots eroded by water during high flows.	9	Frequent or large, causing sediment nearly yearlong OR imminent danger of same.	12
3	Essentially absent from immediate channel area.	2	Present but mostly small twigs and limbs.	4	Present; volume and size are both increasing.	6	Moderate to heavy amounts, predominantly larger sizes.	8

(continued on next page)

Table A-2 (continued)

STABILITY INDICATOR NUMBERS	RATING							
	EXCELLENT		**GOOD**		**FAIR**		**POOR**	
4	>90% plant density. Vigor and variety suggest a deep, dense soil binding root mass.	3	70–90% density. Fewer plant species or lower vigor suggests a less dense or deep root mass.	6	50–70% density. Lower vigor, and still fewer species form a somewhat shallow and discontinuous root mass.	9	<50% density plus fewer species and less vigor indicate poor, discontinuous, and shallow root mass.	12
5	Ample for present plus some increases. Peak flows contained. Width-to-depth ratio <7.	1	Adequate. Overbank flows rare. Width-to-depth ratio 8–15.	2	Barely contains present peaks. Occasional overbank floods. Width-to-depth ratio 15–25.	3	Inadequate. Overbank flows common. Width-to-depth ratio >25.	4
6	>65% rocks in bank with large angular boulders >12″ numerous.	2	40–65%, mostly small boulders to cobbles 6–12″.	4	20–40%, with most in the 3–6″ diameter class.	6	<20% rock fragments of gravel sizes 1–3″ or less.	8
7	Rocks and old logs firmly embedded. Flow pattern without cutting or deposition. Pools and riffles stable.	2	Some present, causing erosive cross currents minor pool filling. Obstructions and deflectors newer and less firm.	4	Moderately frequent, moderately unstable obstructions and deflectors move with high water causing bank cutting and filling of pools.	6	Frequent obstructions and deflectors cause bank erosion yearlong. Sediment traps full, channel migration occurring.	8
8	Little or none evident. Infrequent raw banks <6″ high generally.	4	Some, intermittently at outcurves and constrictions. Raw banks may be up to 12″.	6	Significant. Cuts 12–24″ high. Root mat overhangs and sloughing evident.	12	Almost continuous cuts, some >24″ high. Failure of overhangs frequent.	16
9	Little or no enlargement of channel or point bars.	4	Some new increase in bar formation, mostly from coarse gravels.	8	Moderate deposition of new gravel and course sand on old and some new bars.	12	Extensive deposition of predominantly fine particles. Accelerated bar development.	16

(continued on next page)

STABILITY INDICATOR NUMBERS	RATING			
	EXCELLENT	**GOOD**	**FAIR**	**POOR**
10	Sharp edges and corners, plane surfaces roughened. — 1	Rounded corners and edges, surfaces smooth and flat. — 2	Corners and edges well rounded in two dimensions. — 3	Well rounded in all dimensions, surfaces smooth. — 4
11	Surfaces dull, darkened, or stained. Generally not bright. — 1	Mostly dull, but may have up to 35% bright surfaces. — 2	Generally 50–50% mixture of dull and bright. May be 35–65% dull. — 3	Predominantly bright, >65% exposed or scoured surfaces. — 4
12	Assorted sizes tightly packed or overlapping. — 2	Moderately packed with some overlapping. — 4	Mostly a loose assortment with no apparent overlap. — 6	No packing evident. Loose assortment, easily moved. — 8
13	No change in sizes evident. Stable materials 80–100%. — 4	Distribution shift slight. Stable materials 50–80%. — 8	Moderate change in sizes. Stable materials 20–50%. — 12	Marked distribution change. Stable materials 0–20%. — 16
14	<5% of the bottom affected by scouring and deposition. — 6	5–30% affected. Scour at constrictions and where grades steepen. Some deposition in pools. — 2	30–50% of the bottom affected. Deposits and scour at obstructions, constrictions and bends. Some filling of pools. — 18	>50% of the bottom in a state of flux or change nearly yearlong. — 24
15	Abundant. Growth largely moss-like, dark green, perennial. In swift water too. — 1	Common. Algal forms in low velocity and pool areas. Moss here too, and swifter waters. — 2	Present but spotty, mostly in backwater areas. Seasonal blooms make rocks slick. — 3	Perennial types scarce or absent. Yellow-green, short-term bloom may be present. — 4
	Total Score for Excellent Column ____	**Total Score for Good Column ____**	**Total Score for Fair Column ____**	**Total Score for Poor Column ____**

[1] This scoring sheet is used to determine the stability of the stream channel based on fifteen indicators that are scored excellent, good, fair, or poor. The numbers in the far left column represent stability indicators. Use the Key to Stability Indicators for more information. Circle the score to the right of the description that best matches the stream site for each indicator. After evaluating all of the stability indicators, add up the scores for each column: excellent, good, fair, and poor. Transfer this information to the Stream Channel Stability Assessment Data Form to determine the total score and rating for the site.

Appendix D

Safety and Fun in the Watershed

*T*here are several important things to remember when you are working outside. If you follow these safety tips, you will have a fun and enjoyable experience.

Before You Go

Remember to tell a friend or relative the date, time, and location of your watershed activity. Work with a partner so someone can go for help if you are injured.

Find the phone number and location of the medical center nearest to your work site. Carry a cellular phone with you and note the location of the nearest pay phone. Remember that cell phones do not always work in rural areas, so do not rely on them at all times. Satellite phones, which work anywhere, are more reliable.

Bring a first aid kit that includes these items:

• Adhesive and cloth bandages;

• Antiseptic spray or ointments;

• Surgical tape;

• Hydrogen peroxide;

• Tweezers;

• Cotton balls;

• Aspirin or non-aspirin pain reliever;

• Bee-sting neutralizers.

Review safety rules and tips with everyone in your work group before each outdoor project.

Safety Rules

The Izaak Walton League recommends that groups never get into a stream when the water is at flood stage or is flowing much swifter than normal. It is better to delay monitoring or cleanup projects than to risk personal harm. Water should always be below knee level. Remember that the knee level of children may be much lower than the knee level of adults. Avoid steep and slippery banks.

When in contact with water, keep your hands away from your eyes and mouth. Always wash your hands thoroughly with soap and water after being in contact with stream or river water. You may also want to bring antibacterial hand gel to the field site for use immediately after water contact.

If the water is posted as unsafe for human contact or appears to be severely polluted (e.g., strong odor of sewage or chemicals, unusual colors, lots of dead fish), **do not touch the water**. If these signs of severe pollution are not present but you are unsure of conditions or would like additional protection, take the following precautions:

- Wear rubber boots high enough to keep water from contacting your skin.
- Wear heavy rubber gloves that go up to your shoulders (available at most automotive supply stores). Surgical gloves will not work. They can be punctured easily by snags or sharp objects, and they are not long enough to protect your arms.
- Wear a protective covering for your mouth such as a painter's mask (available at most drugstores or hardware stores). You can get sick if you breathe in vapors from sewage-contaminated water.
- Report any pollution problems to your state's water regulatory agency.

Other Areas of Concern

Snakes

Snakes can be a concern when you are in an aquatic environment, especially slow-moving waters with overhanging vegetation. To avoid an encounter with a snake, observe the following rules:

Check rocks, logs, and stubs for snakes. Snakes must get out of the water to dry their skin and will lie on flat surfaces exposed to sunlight.

If you have to approach the water through high grass, thump the ground in front of you with a stick. Snakes will feel the vibrations and move away. Snakes are deaf and respond only to vibrations.

If you come upon a snake at close range, simply move away. The snake probably will leave the area when it no longer perceives you as a threat. Remember, you are much bigger than the snake, and it is more afraid of you than you are of it. Allow the snake a chance to back off, and it usually will.

Most snakes associated with aquatic environments are not poisonous. However, because it's difficult to distinguish between poisonous and non-poisonous snakes without getting too close, the best advice is to stay away from all of them. If a snake does bite, follow these simple steps:

- Remain calm. Take a few deep breaths and keep movement to a minimum.
- Elevate the bitten area. Do not apply ice or a tourniquet to the wound. Do not cut the wound open or attempt to suck out the venom.
- Remove all watches and jewelry on the hand or arm if near the bite. Snake venom will cause the bitten area to swell.
- Seek immediate medical attention. Walk calmly to your vehicle and have your partner carry your equipment.

Insects

If you are allergic to any types of insects, bring your antidotes or medicines. Ask other members of your group about their allergies before you go to the site. If a

volunteer gets an insect bite that swells up to an unusual size or has severe redness, seek medical attention immediately.

Many people have concerns about West Nile Virus. Female mosquitoes transmit the virus primarily to birds. Occasionally, mosquitoes transfer the virus from birds to humans, most of whom experience no symptoms. About one in five infected people develop West Nile fever, which resembles the flu.

Infections can be fatal in people with weak immune systems, but this is rare. To avoid mosquito bites, wear long sleeves and pants. Avoid areas of standing water during dawn and dusk, when mosquito activity is at its peak. Consider using mosquito repellants that contain DEET. Do not spray DEET underneath clothes. For more information on West Nile Virus, see the Environmental Protection Agency factsheet *Wetlands and West Nile Virus* on-line at *www.epa.gov/owow/wetlands/facts/WestNile.pdf,* or contact the Izaak Walton League.

Ticks

Ticks are prevalent in grassy or woody areas. It is important for volunteers to check their bodies for ticks. Feel along the scalp for any loosely attached bumps. If it is a tick, do not pull it out. Yanking the tick may cause an infection if the tick's head remains in the scalp. Grasp the tick with tweezers and gently twist it counterclockwise for several rotations until the tick is free. Swab the area with hydrogen peroxide to clean the area. If you want to kill the tick, burn it with a match or suffocate it with nail polish or petroleum jelly after it has been removed from the skin.

One type of tick, called a deer tick, can carry a serious illness called Lyme Disease. Deer ticks resemble common ticks except they are much smaller (only a few millimeters across). Symptoms of Lyme Disease include chills, malaise, and fever. The first sign of Lyme Disease is often a bull's-eye shaped mark on the skin, but this is not always present. Treatment requires an injection of prescribed antibiotics. If not treated, this disease can remain in your body for a lifetime. If you exhibit any of the symptoms, it is recommended that you see your doctor and ask for a Lyme Disease test.

Alligators and Turtles

In southern states, you may encounter alligators and large aquatic turtles. These animals are not dangerous if left alone. Alligators under 18 inches in length are juveniles and may be near their mothers. Female alligators are very protective and may be dangerous. If you see alligators, leave the area immediately. Snapping turtles and soft-shelled turtles usually will move out of an area if the water is disturbed. Although turtles are not poisonous, treat a turtle bite with the same care as a snake bite.

Bears

Black bears and grizzly bears live in forested areas around the United States. Black bear encounters are more prevalent in the eastern US, while grizzlies may be encountered in the Northwest.

- When in an area with the potential for bear encounters, make sure you stay with a group of people and make noise to alert the bears of your presence.
- If you see a bear and it does not see you, quickly leave the area while keeping your distance from the bear, giving it plenty of room to escape should you startle it.

- If you encounter a bear and it sees you, do not run. You cannot outrun a bear. Stay calm and slowly back away from the bear. Look for an escape route that gives the bear plenty of space; try to stay out of its "comfort zone."

- Climbing trees to escape is a common suggestion, but be aware that many bears can follow you up a tree.

- If a bear should charge you, do not run. Drop to the ground and cover your head, face, and neck with your arms for protection. If you are wearing a backpack, make sure it faces the bear so it can absorb punishment from any attack. Bear attacks are often "hit and run" and do not last very long. Lay motionless and give the bear time to leave the area. Seek medical treatment as soon as possible for any injuries.

- If you feel an attack is predatory, disregard the above strategy and fight back with everything you have. This also applies to mountain lion attacks. Seek medical treatment immediately and report the attack to wildlife authorities.

- Never go near a cub because the mother bear is always nearby and will become very aggressive in trying to protect her young.

Appendix E

Establishing a Mission and Goals

*T*he creation of mission and goal statements is critical to project success. In the hierarchy of organizational planning, a mission statement defines the overall purpose of an organization, goals define the specific outcomes expected, and objectives describe how to achieve the goals. It is hard to state the goals without a clearly defined mission (Bolling, 1994). A mission, whether broad or specific, helps to provide focus.

For example, the Izaak Walton League's mission is:

> *To conserve, maintain, protect, and restore the soil, forest, water and other natural resources of the United States and other lands; to promote means and opportunities for the education of the public with respect to such resources and their enjoyment and wholesome utilization.*

The League's Save Our Streams program was created to fulfill the goal of stream stewardship, and this handbook is one of several tools to support the program's goals. If the general purpose or mission of an effort is to enhance a local stream corridor, the goal could be to "Enhance riparian conditions to improve water quality, wildlife and fish habitat, and scenic value."

Goals based on the overall mission help both to motivate volunteers and evaluate program success. It is important to reevaluate goals and adapt them to the changing needs and resources of the program. Outline steps needed to accomplish your goals and delegate the objectives. Goals can be short-term or long-term, depending on the nature of the project. Determining if goals require additional funding or community support is also critical. Consider budget and timeline restrictions in determining project objectives. Realistic goals reflect the timeline, commitment level, budget, and actions required to fulfill the community's needs.

Some goals may extend throughout the life of the program. For example, if one of the goals is to educate every resident in the community about good land-use practices, new landowners need to be educated as they move into the community. In northern Virginia, near Washington, DC, half of the population moves every five years. That means education programs must be constant to keep new community members informed and involved. Even in more stable regions of the country, a mechanism is needed for periodically reaching out to community members to keep them informed about the program and its successes. Sending out a periodic newsletter is one way of keeping the community informed. Build a relationship with your local newspaper and always invite the press to important meetings and field days. Send invitations to elected officials and remember to notify the public about the project's progress. Media attention can help recruit new volunteers and partners.

Consider some or all of the following goals in a stream stewardship program.

Shorter-Term Goals

- Hold an "Introduction to Stream Bank Buffers" workshop and register 15 volunteers to be stream-bank stewards.
- Improve stream bank habitat through a trash cleanup and a stream-bank planting project.
- Erect a sign along the stream bank explaining your stewardship program. If you live in a diverse neighborhood, print the sign in different languages.
- Establish a "Stream Stewardship Day" in your community to focus attention on the importance of stream banks.
- Organize volunteers for a project to stabilize eroding stream banks.

- Install vegetative plantings along stream banks to improve water quality and riparian habitat.

Longer-Term Goals

- Develop a volunteer monitoring strategy to provide baseline data about the health of the stream.
- Identify priority sites for enhancement in the watershed.
- Develop an enhancement strategy for the watershed.
- Create a long-term stewardship program for the stream that includes educating every seventh-grader in the county about stream ecology.
- Create a stream bank community center to expand recreational opportunities along the stream.
- Plan for bike paths or nature trails.
- Gather data to be used in local land-use decisions.
- Build an interpretive boardwalk along the stream.
- Design a community park along the stream.
- Produce an annual report detailing your monitoring efforts and distribute it to local government officials, the media, etc.

Bringing people to the stream and creating more opportunities for enjoyment and recreation may be the most important way to get participation from diverse interests. People are more inclined to take care of environments that are fun and where they feel safe, comfortable, and relaxed. Stream bank habitats can inspire local artists to portray the community as a nice place to live and can encourage ecotourism and economic growth. Continue to reevaluate your program's goals to reflect the community's needs.

Appendix F

Channel Cross-Section Data Form and Instructions

Measuring the channel cross-section can help determine the shape and type of stream at a particular area. Channel cross-section measurements also are necessary to determine the bankfull or channel-forming discharge. Measurements can be taken at the same location over a period of time and may indicate areas of erosion and deposition. This appendix provides instructions for measuring channel cross-section. An example of a data set and its graphic representation (Figure A-6) and blank forms for recording and plotting data (Figure A-7) are provided on the following pages.

Instructions for Measuring Channel Cross-Section

- Select an area to take the cross-section data that is in a straight reach of stream (not a bend or riffle).
- Drive a piece of rebar into the top of each bank. One rebar should have a clip to hold the measuring tape and the other should have a hook for the tape and the nylon line.
- Attach line and measuring tape to the rebar with the hook. Stretch the line and measuring tape across the stream to the other rebar. Pull the line and measuring tape taut and secure them with the clip.
- Check to make sure the line is level. Hook the line level onto the line in the middle of the stream channel. Adjust the height of one rebar until the air bubble on the line level is between the two black lines.
- Using a surveyor's rod or tall pole marked with feet and inches, measure the height of the line above different points across the stream channel. Starting at the first rebar, record the height of the nylon line above the top of the bank. Continue to take a similar vertical measurement at every horizon-

tal foot across the channel. Take additional measurements whenever there is a steep elevation change on the bank or in the channel. Also take a measurement at the water line and the bankfull stage if the bankfull stage can be determined. Record each horizontal distance and its corresponding vertical elevation as shown in the example below.

• The vertical distance measured is the difference in elevation between the nylon line and the ground. Convert these vertical distance measurements into actual elevations above sea level by determining the nylon line's true elevation above sea level. (For example, if the elevation of the line is 100 feet, subtract each vertical measurement from 100 to obtain the actual elevations of the points being measured.)

• Plot the horizontal distances on the x-axis (horizontal) and the corresponding elevation on the y-axis (vertical).

SAMPLE CHANNEL CROSS-SECTION DATA FORM

HORIZONTAL DISTANCE	VERTICAL DISTANCE	ELEVATION (elevation of line minus vertical distance)
0 ft.	3 ft.	(100-3) = 97 ft.
1 ft.	3 ft.	(100-3) = 97 ft.
2 ft.	3.2 ft.	(100-3.2) = 96.8 ft.
2.5 ft.	5.5 ft.	(100-5.5) = 94.5 ft.
2.8 ft.	7 ft.	(100-7) = 93 ft.
3 ft.	9 ft.	(100-9) = 91 ft.
4 ft.	9 ft.	(100-9) = 91 ft.
5 ft.	8.8 ft.	(100-8.8) = 91.2 ft.
5.2 ft.	8 ft.	(100-8) = 92 ft.
6 ft.	7.9 ft.	(100-7.9) = 92.1 ft.
7 ft.	7.8 ft.	(100-7.8) = 92.2 ft.
8 ft.	6 ft.	(100-6) = 94 ft.
9 ft.	3.5 ft.	(100-3.5) - 96.5 ft.
10 ft.	3 ft.	(100-3) = 97 ft.

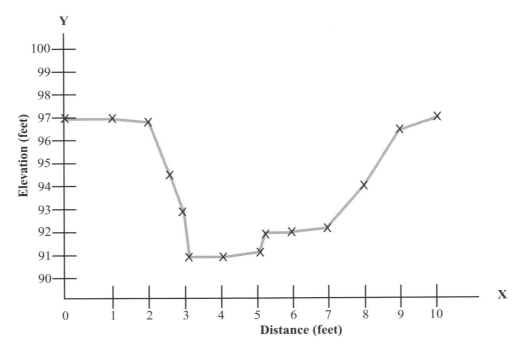

Figure A-6. (Top) Channel cross-section data that is collected using the techniques described in this Appendix can be recorded in this format. (Bottom) Channel cross-section data can be converted to a drawing that shows the shape of the stream channel using the techniques described in this Appendix.

CHANNEL CROSS-SECTION DATA FORM EVALUATING STREAM ENHANCEMENT SUCCESS

HORIZONTAL DISTANCE	VERTICAL DISTANCE	ELEVATION (elevation of line minus vertical distance)

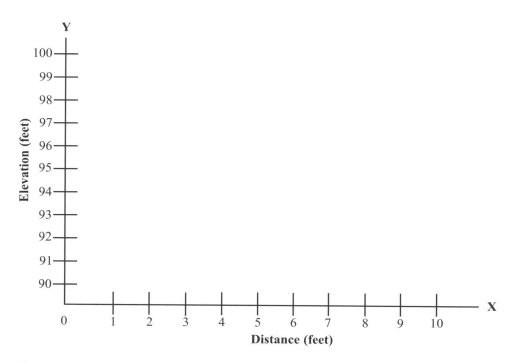

Figure A-7. (Top) This form can be used to collect channel cross-section data in the field. (Bottom) This form can be used to transfer channel cross-section data collected in the field to a drawing of the stream channel.

Appendix G

Evaluating Stream Enhancement Success

*T*he following table will help you choose monitoring techniques to evaluate the success of a project in meeting specific objectives (Table A-3).

Table A-3. Matching Monitoring Techniques to Project Objectives.

GENERAL PROJECT OBJECTIVES	EVALUATION/ MONITORING TECHNIQUE	DESCRIPTION
Stabilize channel and banks; stop erosion	Channel cross-sections	Shape of channel changes recorded over time can indicate areas of erosion and deposition and helps determine stream type. This allows for management based on stream type and conditions.
	Bank and bed pins	Can be used to measure the rates of bank erosion over time by driving pins into side of the bank and bed and checking at set time intervals for buried or uncovered pins.
	Photo points	Take photographs of stream channel from set locations to visually compare changes in channel shape over time.
	Width-to-depth ratio	The average bankfull width over the average bankfull depth. Defines the distribution of enegy of flow through the channel. As width/depth ratios increase, the channel becomes relatively wider and more shallow, water temperature at low flow increases, aquatic habitat decreases, and the channel's capability to transport sediment is reduced.

(continued on next page)

Table A-3 (continued)

GENERAL PROJECT OBJECTIVES	EVALUATION/ MONITORING TECHNIQUE	DESCRIPTION
Stabilize channel and banks; stop erosion (continued)	Entrenchment ratio	Index value used to describe the degree of vertical containmnent of a river channel. An entrenched channel holds floodwaters within the channel causing erosion.
	Bank slope	Change in vertical distance over change in horizontal distance from top of bank to bankfull. Steep slopes are less stable. A 3:1 or a 4:1 slope is preferred for vegetation to grow successfully on banks.
	Bank erosion	Indicates the ability of streambanks to resist potential rating erosion based on several factors such as the composition of streambank materials, bank slope, riparian vegetation rooting density, and ratio of streambank height to bankfull stage.
	Aerial photography	Shows changes in overall channel shape over time; can be expensive and may need expert help to interpret.
Improve aquatic habitat	Macroinvertebrate monitoring	A diversity of aquatic insects and crustaceans indicates that water quality, physical characteristics of the stream, and streamside vegetation are healthy. Macroinvertebrates provide food for other aquatic organisms, like fish.
	Fish monitoring	Especially useful if restoring fish habitat is a goal of the project.
	Percentage overhang, shading of channel by vegetation	Shading of channel keeps stream temperatures down which increases the level of dissolved oxygen in water.
	Pool/riffle composition	Pools and riffles are important habitat for fish and aquatic macroinvertebrates.
	Stream temperature	Lower temperatures are required for aquatic life and allow greater levels of dissolved oxygen.
Improve water quality[a]	Chemical monitoring	Turbidity, pH, dissolved oxygen, temperature, solids, trash, nitrates, toxics — depending on the type of restoration you do, only some of these may be affected.
	Macroinvertebrate and fish monitoring	Different organisms have different tolerances to pollution. Water quality ratings have been developed based on the presence or absence of pollution sensitive species as well as on overall species diversity.

(continued on next page)

Table A-3 (continued)

GENERAL PROJECT OBJECTIVES	EVALUATION/ MONITORING TECHNIQUE	DESCRIPTION
Improve riparian habitat	Percentage vegetative cover	Evaluate the wildlife of the riparian corridor by measuring the percent cover of each species along a transect. This information can be used to keep track of invasive exotic species.
	Planting survival	Count trees, shrubs, and other vegetation planted on a site over time to see how many of each species survive.
	Width of buffer	Estimate average width of vegetative buffer between stream to developed areas.
	Vegetative species diversity	Qualitative or quantitative analysis of the diversity of vegetation (presence of trees, shrubs, vines and ground cover; number of different species; exotic vs. native species).
	Photo points	Take photographs at permanent monitoring locations and compare vegetation cover over time.
	Aerial photography	Can be used to evaluate overall vegetation cover of the project site. Can be expensive.
Community involvement and education	Community participation in project	Keep track of the number of community members who participate in the project.
	Workshop evaluations, community surveys	If holding an educational workshop, evaluations help keep track of what people learned. Community surveys can evaluate the success of more general educational efforts within the community.

[a] The project team should isolate the effects of the project on water quality by sampling upstream and downstream of the project site both before and after the project takes place.

A Handbook for Stream Enhancement & Stewardship

Appendix H

Land-Use Planning

Establishing riparian buffers and installing bioengineering techniques to stabilize stream corridors, create habitat, and intercept contaminants from surface and groundwater before they reach a stream are most effective when combined with a sound land-management plan. Land uses affect the quality and quantity of water reaching streams and other water bodies. The land-use planning process incorporates several opportunities for public participation in decision making.

Each township, city, or county usually has a land-use plan, called the master or comprehensive plan. If a town does not, it's probably time to develop one. This plan is a guide for where homes, businesses, and parks are to be developed. A community's plan can be found at the planning or zoning office, at the planning commission, or in the community's public library.

Communities often have other legal means for managing growth and development, such as zoning ordinances and subdivision regulations. Zoning, for ex-

ample, sets rules for the location of land uses, how many houses per acre are allowed, how tall buildings can be, and other restrictions. Subdivision rules state how far back from the street houses must be, the amount of parking, and the width of streets. Current versions of many of these ordinances and plans contain outdated standards.

There are many opportunities for citizen groups to get involved with local land-use planning. Citizens can attend planning commission meetings. At these meetings, the planning commission makes recommendations to the county or town council or commissioners concerning changes in land-use plans or zoning and takes requests by developers for particular projects. Citizens also can participate in the process for reviewing and making changes to the local land-use plan or zoning ordinance, which occurs every few years.

Expert land-use planners agree that compact development, within or immediately adjacent to the urban area, is an important first step in managing growth.

Land use directly affects the environment and is important to the quality of our everyday lives. State governments often give local governments broad authority for land-use planning and zoning, with clear environmental and other objectives. Local governments can help to focus growth in areas that have already invested public money in roads, sewers, and other infrastructure and make rural countryside off-limits to growth. Some counties are creating programs to buy development rights from landowners to preserve land. Cities, towns, and villages should be zoned to attract economic and residential growth, mixing land uses and densities. Rural land should be zoned to keep open space undeveloped.

Stormwater controls also protect stream health and need community support to be maintained. Every day, streets, driveways, parking lots, rooftops, and other hard surfaces collect nutrients and chemicals. They include waste from pets, air pollutants that fall from the sky, oils and antifreeze from cars, waste dumped by residents, excess fertilizer applied to lawns, and many other substances. Every time it rains, these byproducts of everyday life are collected and swept down the storm drains, ultimately pouring into the local stream. The runoff also accelerates as gravity pulls it down through the pipes and it often hits the streams at high velocity, gouging sediment from the banks and smothering wildlife in the stream.

Stream degradation resulting from development is on the rise since pavement prevents rainwater from percolating through the soil, a process that removes pollutants and recharges groundwater. Ordinances, like no-mow zones around streams, ensure that a buffer exists to filter runoff as it moves toward a stream. Citizens can promote low-impact development that emphasizes techniques to capture rainwater throughout a development and allow it to infiltrate into the ground. Rain gardens, vegetative swales, and pervious pavement all allow water to seep through. It is important to support developers and local government planners as they look for ways impervious surfaces can be reduced, thereby reducing the amount of runoff. Stormwater retention ponds are another technique included in development designs to capture and hold runoff while it seeps into the ground.

Citizens can reduce their impacts by becoming more aware of what may be causing stormwater pollution in the area. Citizens should also report dumping of inappropriate materials (such as oil and antifreeze) into stormwater drains and construction sites that do not have erosion or sediment controls. Residents can be good watershed stewards with their use of lawn-care chemicals, oil, gasoline, pet waste, etc. Communities also can start or participate in programs to recycle and safely dispose of used oil and household hazardous wastes and containers. As mentioned earlier, citizens can attend meetings of the planning commission to support changes to stormwater management required for construction permits. Citizens can also recommend that the local government mandate the maintenance of stormwater systems.

For more information on low impact development, please visit *www.lid-stormwater.net,* and *www.lowimpactdevelopment.org*. Both of these sites provide tools and techniques to incorporate low-impact development into stream enhancement plans. The Low Impact Development Center (*www.lowimpactdevelopment.org*) can provide research, training, planning, development, design, and monitoring assistance.

Appendix I

Obtaining a Permit

State and federal permits might be required for enhancement projects that alter stream corridors. Many of the structural techniques that involve moving soil require a permit. Even planting vegetation on stream banks could require a permit if grading is necessary. Usually, stabilization projects that involve work below the ordinary high-water mark and closer to the streambed require stricter permits than those above this mark. The Army Corps of Engineers interprets the ordinary high-water mark to be synonymous with the bankfull discharge water level. Unfortunately, the permitting process is not always simple. Permits may be needed from several state and federal agencies, and requirements vary from state to state.

Approval for permits can take months, and waiting until the last minute to apply for one could delay work plans. If the project is time sensitive (e.g., installation must take place in the low-flow season or in late winter, when the plants are dormant), apply for a permit well ahead of the time that it is needed. If consulted early on, the permitting agencies might be able to offer helpful suggestions and design alternatives for the project.

The Army Corps of Engineers regulates permitting under Section 404 of the Clean Water Act. The Corps issues permits for dredging or filling of, installations in, or alterations to "the waters of the United States." The Corps separates its permits into nationwide and individual permits. Nationwide permits are issued for many activities considered not detrimental to the environment. These permits outline criteria for specific activities including:

- Bank stabilization activities necessary for erosion protection;
- Removal of temporary structures or man-made obstructions to navigation;
- Riparian enhancement and creation activities, including installation and

maintenance of small water-control structures such as log weirs, dikes, and berms;

• Backfilling of drainage ditches;

• Removal of drainage structures;

• Construction of small nesting islands; and

• Other related activities.

The Corps has division and district offices located across the country. Find a district office for guidance on the permitting process by calling the Corps at (202) 761-0010. If the office says you will need a permit, see if your project can be covered under an existing general permit. An activity not covered by a general permit will be considered for an individual permit. Individual permits require a more formal and lengthy application process that includes meetings with representatives from state and federal agencies.

Any activity affecting your state's waterways must meet state water quality standards under Section 401 of the Clean Water Act. Many states require a stream alteration permit for any work done within the stream channel. They may also require permits addressing water quality or other environmental impacts. Contact the state environmental protection office to get information on these permits or guidance on how to apply.

A project also might fall under the purview of state regulatory agencies if it is located within a designated coastal management area, critical habitat area, or other special land designation. Some states require Aquatic Resource Alteration Permits for projects adjacent to public lands. Many counties have ordinances that are more stringent than state laws. Contact local or county planning boards directly to learn if any local level permits are required. The local government engineering or code enforcement office might also have a role in overseeing local regulations.

Projects need to be completely planned before applying for permits, and work may not begin until state and federal approval is received. The following information may be required for permit applications:

• Applicant name and address;

• Purpose of project;

• Location and size of project;

• Names of adjacent landowners;

• Installation procedures;

• Type of material to be installed;

• Drawings of existing conditions and conceptual plans, illustrated in-stream dimension, pattern, and profile views.

After reviewing your permit application, the agencies might respond in three ways: 1) No further state authorization is required and the permit is granted; 2) the permit is granted with specific stipulations to lessen impacts on the environment; or 3) the permit is denied.

Most aquatic habitat enhancement projects require a permit, but the process is relatively simple because the project will benefit the environment. These permits are designed to keep the proper authorities aware of alterations to streams and their watersheds so they can protect streams from degradation. No one is exempt, and there may be substantial civil or criminal liability for those who ignore the need for a permit.

Three sample drawings of a stabilization design submitted for permit review follow (figures A-8, A-9, A-10).

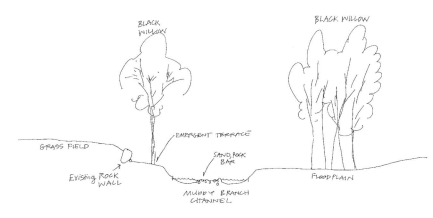

BLACK WILLOW

BLACK WILLOW

GRASS FIELD

EMERGENT TERRACE

SAND, ROCK BAR

Existing ROCK WALL

FLOODPLAIN

MUDDY BRANCH CHANNEL

PHASE ONE · MUDDY BRANCH ENHANCEMENT

EXISTING CONDITIONS

CROSS-SECTION A-A'

Figure A-8. A drawing of the existing conditions along a stream corridor can establish the basis for proceeding with an enhancement project and is required when submitting an application for permit review. (Izaak Walton League of America figure)

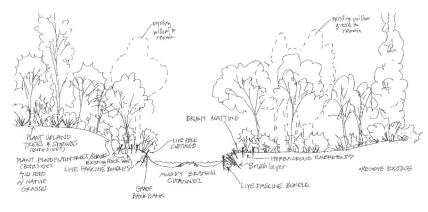

existing willow to remain

existing willow grove to remain

BRUSH MATTING

LIVE POLE CUTTINGS

PLANT UPLAND TREES & SHRUBS (BOTH SIDES)

PLANT FLOODPLAIN TREES/SHRUBS Existing Rock Wall CBOTH SIDES) LIVE FASCINE BUNDLES AND SEED W NATIVE GRASSES

GRADE BACK BANK

MUDDY BRANCH CHANNEL

Brushlayer

LIVE FASCINE BUNDLE

HERBACEOUS EMERGENTS

•REMOVE EXOTICS

PHASE ONE MUDDY BRANCH ENHANCEMENT

CONCEPTUAL PLAN

CROSS-SECTION A-A'

Figure A-9. A cross-section sketch of a proposed enhancement project can clearly depict the design of planned stream bank stabilization techniques. In addition to making sure that the project is built according to the design, this sketch is required by permit review agencies. (Izaak Walton League of America figure)

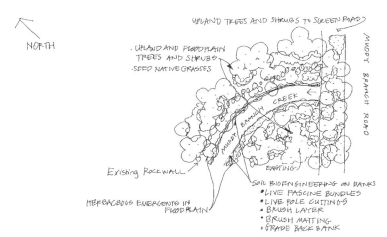

NORTH

UPLAND TREES AND SHRUBS TO SCREEN ROAD

· UPLAND AND FLOODPLAIN TREES AND SHRUBS ·SEED NATIVE GRASSES

MUDDY BRANCH ROAD

MUDDY BRANCH CREEK

Existing ROCKWALL

HERBACEOUS EMERGENTS IN FLOODPLAIN

EXISTING

SOIL BIOENGINEERING ON BANKS
•LIVE FASCINE BUNDLES
•LIVE POLE CUTTINGS
• BRUSH LAYER
• BRUSH MATTING
• GRADE BACK BANK

PHASE ONE · MUDDY BRANCH ENHANCEMENT

CONCEPTUAL PLAN

PLAN VIEW

Figure A-10. A plan view of the project proposal can better portray the anticipated changes in the surrounding landscape and is also required in the permit review process. (Izaak Walton League of America figure)

Funding Watershed Conservation Projects

*A*ccomplishing your individual or group watershed conservation goals often requires soliciting resources from outside funders. Raising money is time consuming and requires hard work, but the reward is that you will be able to achieve your goals of conserving your watershed.

First, it is important to have a fundraising plan. Determine the amount of money or types of funding sources that are needed to complete your project or fulfill your group's goals. Develop a budget that shows the amount of money needed for equipment, telephone, office space, printing, postage, office supplies, consultants or contractors, salaries (if paid staff is involved), etc. This budget will help you to calculate how much money you need and to draft a wish list for in-kind donations.

In-Kind Donations

No matter how large or small your project, seek in-kind donations. The more materials, printing, tools, refreshments, meeting space, and professional services that are donated, the less money needs to be raised. In-kind gifts also can be used to leverage grant funds. Many funders require matching funds from other donors, including in-kind donors. Remember to ask the donor for a letter that states the value of the in-kind contribution.

Let members of your group know about your wish list and talk about which businesses to approach. Group members might have personal connections with businesses. Also, make your in-kind donation needs known by publishing a wish list in your newsletter or on your web site. Call or visit local businesses to request in-kind donations. Many donors will want a written request, so follow up with a letter. Be sure to start this process early and try to target at least three sources for every in-kind item that you need for your project. Finally, give the donors a positive giving experience by thanking them personally and publicly.

Encourage your members and partners to patronize businesses that donated the in-kind gifts to your project.

Finding Grant-Makers

In addition to in-kind donations, you may want to seek grants or donations. The first step in this process is finding potential funders. Government agencies, some nonprofit organizations, foundations, and corporations are all potential sources of grant funds. Most grantors require proof of 501(c)(3) nonprofit status. If your group is not incorporated as a nonprofit with a tax classification assigned by the Internal Revenue Service, you may want to consider incorporating and applying for nonprofit status. One way of doing this is by becoming a chapter of the Izaak Walton League and being under the League's nonprofit status. However, your group may also find a partner organization with 501(c)(3) status that may be willing to pass grant money through to your group as a subcontractor.

Start locally. Many foundations and corporations contribute only to local organizations. Look for foundations and corporations in your state. Finally, tap regional and national funders.

There are many directories and other resources to help groups identify foundations and corporations for fundraising. Look for a statewide foundation directory at your local library or purchase a copy for your use. You also can ask a larger nonprofit or a local foundation to donate last year's edition of a regional or national directory. There are on-line resources as well. Many on-line directories and data-bases for funders charge a fee, but there are also free resources. You can locate prospects by searching the Internet. Other sources of information about potential

funders are the annual reports of other organizations that have similar programs. Read these reports to see which foundations, agencies, and corporations are supporting your type of project.

It is important to target your solicitations. Sending out dozens of unsolicited proposals to foundations without first reading their guidelines is unlikely to yield results. When researching grant-makers, pay attention to geographic restrictions, types of projects funded, and the size of grants. Obtain the most up-to-date grant guidelines, annual reports, and brochures by contacting the foundation directly or visiting its web site. In addition to grant guidelines, pay close attention to the types of projects the foundation, corporation, or agency has previously funded. For example, a foundation may list water quality as an interest, but a review of past grants may show that it gives primarily to university research rather than volunteer monitoring projects. The list of past grantors will also provide information on the range of funding you might request.

Writing Grant Proposals

Before writing a proposal, it is wise to contact the funder to find out whether or not they have grant guidelines, what the deadlines are for submittal, and whether or not your project is within the funder's area of interest. If the grant guidelines specify that the foundation does not accept tele-phone calls, remember to honor that request.

Start with a thorough outline of the project. Think about why the project is needed and make that clear. It is important to keep asking yourself whether this particular project or program is the best thing you can do to solve the problem you are addressing. Think about the project

goals, work plan, time frame, and budget. This information will work well for either a brief letter of inquiry or a full proposal. If submission guidelines are available, make sure you follow them with regard to length, format, and content. Tailor each proposal to the particular funder. Review information on the funder carefully and get a sense of its philosophy and interests. Keep the proposal short, concise, and reader-friendly. Use short, powerful sentences and a logical structure. Proof-read the proposal carefully and ask others to review it. Be sure to allow sufficient time to complete the proposal and submit it by the funder's deadline.

After sending the proposal, follow up with the funder. If your project is approved, send a thank-you letter immediately. If your project is not funded, you may want to call and inquire why the funds were denied.

Remember that people give money to people. Developing relationships with grant-makers is very important. A good way to establish relationships with funders is to meet them in person. Send a letter requesting a meeting with the funder and follow up with a phone call. Also, invite the funder to your next event to observe your group in action.

Resources

Catalog of Federal Funding Sources for Watershed Protection. This comprehensive listing of federal funding available for watershed projects includes detailed information on each funding source and links for more information. Call (800) 490-9198 or (513) 489-8695 or visit *www.epa.gov/OWOW/watershed/wacademy/.*

The Chronicle of Philanthropy. This publication's web site features a free,

searchable database for all the grants listed in its issues during the past several years. Visit *www.philanthropy.com.*

Environmental Finance Center. This organization features an on-line, searchable database for watershed restoration funding including federal, state, private, and other funding sources for the Pacific Northwest. It also has free software that helps users estimate the costs of their projects and determine funding needs. Visit *sspa.boisestate.edu/efc/Tools&Services/Plan2Fund/plan2fund.htm* or call (866) 627-9847.

The Foundation Center. This center offers free links to grantmaker web sites, including private and public foundations, corporations, and community foundations. This site also includes a short course that teaches the basic elements of writing a good proposal. A comprehensive on-line directory of funders is available for a fee, or you can visit one of their libraries and research all of their information for free. Call (800) 424-9836 or visit *www.fdncenter.org.*

Grants.gov. This site allows organizations to electronically find and apply for competitive grant opportunities from all federal grant-making agencies. Visit *www.grants.gov.*

National Fish and Wildlife Foundation. This private, nonprofit organization provides matching grants for on-the-ground conservation projects through a combination of private and public sources of funding. Visit *www.nfwf.org.*

River Network. This group's web site includes factsheets on foundation research, grant writing, and raising funds through boards, the Internet, bequests, workplace giving, and in-kind donations. Its *Directory of Funding Resources* lists more than 300 private, corporate, and federal funding sources for river and watershed groups. The directory is free for River Network Partners (partnership costs $100), or it is available as a hard copy for $35. Call (800) 423-6747 or visit *www.rivernetwork.org.*

For a comprehensive listing of funding resources, visit the Izaak Walton League's Watershed Stewardship Resources available on-line at *www.iwla.org/ sos/resources.*

Appendix K

Vegetative Cuttings

Vegetative cuttings are live plant materials — twigs and branches — that are cut from plants and then placed in the ground to root and grow. In a wetland, the best plants for vegetative cuttings are deciduous plants that root easily in the ground. Willow and cottonwood are two examples. You can also experiment with cuttings from other deciduous shrubs and trees.

Why Use Vegetative Cuttings?

- They are a cost-effective way to supplement or augment vegetation at your enhancement site, or to use in combination with container plants or bare root plants.
- The plant roots and vegetation help reduce erosion by holding the soil together when it rains.
- Over time, as the cuttings root and grow, they will provide habitat for wildlife.

- They work well in difficult planting situations such as rock slopes, boulder outcroppings, or other areas difficult to access.

When Is It Best to Use Cuttings?

Cuttings are best used where there is a source of easily rooting deciduous plant material in your area. Look for a wetland or riparian area that has a large number of the plant species that you are interested in collecting. Be sure to ask the site owner's permission before doing any cutting. Only collect the amount of plant materials that you need. Leave the site looking as though it has been thinned or pruned, not over harvested.

Preparing a Vegetative Cutting

- Take cuttings near the base of main branches of tree or shrub rather than new growth tips (diameter: 0.5–3" in diameter) (Figure A-11A).

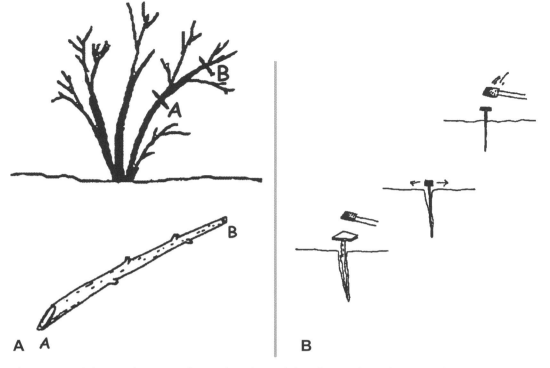

Figure A-11. (A) Preparing vegetative cuttings is a quick and easy alternative to purchasing more costly container and bare-root plants. (B) To ensure that plant cuttings remain firmly anchored in the ground, make the planting hole as deep as possible. (The Wetlands Conservancy, 1996)

- This tip is above ground.
- Remove lateral branches with bark intact.
- Basal (butt) end cut at an angel. This is the end you plant in the soil.

Placing Vegetative Cutting in Ground

- Make a planting hole with a sledge hammer and construction stake (Figure A-11B).
- Make hole as deep as possible, and wiggle construction stake in between sledge hammer blows so stake can be removed easily.
- Place basal (butt) end of cutting in hole. Place a board on top of cutting and hammer a minimum of two-thirds of the cutting in the hole. Fill hole with soil and tamp soil into hole. If cutting splits, discard and place a new one in the hole.

When to Plant

It is best to obtain cuttings when the source plant is dormant, and to install the cuttings soon thereafter. This is usually done during fall and winter. Plant cuttings can be installed in the spring and summer, but may require watering until the fall.

Maintenance

- Cuttings need to be watered during the dry season.
- If cuttings have not rooted and sprouted during the growing season, replace them in the fall with freshly cut vegetative cuttings. If they have come out of the ground, replace them with fresh cuttings.
- Refer to Appendix N for plant protection measures.

[This appendix is reprinted with permission and some modifications from The Wetlands Conservancy, 1996.]

Appendix L

Container Plants

Container plants are nursery stock plants. They are grown in containers that can range in size from four inches in diameter to 15 gallons and larger (Figure A-12A).

Why Use Container Plants?

• Larger plants will provide instant gratification and quickly lay out a foundation that you can continue to develop.

• Larger trees and shrubs provide the canopy and shade needed to establish certain understory plant material.

• They are generally available year round and in a variety of sizes.

• You can buy a container plant in bloom or bearing fruit and see exactly what you are getting.

Where and How to Obtain Container Plants

Container plants can be purchased from retail nurseries and farmers' markets.

If you are planning a large project, it is important to place an order well in advance of your planting dates, to ensure availability. Price will vary based on size and time of year purchased.

Buying Plants

When shopping for container plants, look for plants with a vigorous appearance and healthy foliage.

Avoid:

• Plants with root systems that are badly tangled or knotted.

• Young trees that have roots circling around the trunk.

• Plants with crowded roots. Avoid plants that are too large for the size of the containers, are "leggy," or have dead twigs or branches. Often these are symptoms of plants with crowded roots. If plant roots are exposed on the soil surface or growing through holes at the bottom of the container, this is probably a root-bound plant.

How to Remove Plants from Containers

Plant containers come in a variety of shapes and materials. Each type requires a different technique for plant removal.

Metal Cans

If the plant is in a straight-sided metal can, have the nursery slit down two sides of the can. Be careful not to cut yourself with the metal edges. If you are not going to be planting immediately, you may wait to cut the cans until you are at the planting site.

Tapered or Plastic Containers

If the plant is in a tapered metal can or a plastic container, turn it over and gently tap the plant out of the container.

Fiber Container

If the plant is in a fiber pot, tear the pot away from the root ball.

Planting Container Plants

• Remove plants from the container. Spray soil off the outer few inches of root ball or loosen and uncoil circling or twisted roots by hand. You may need to cut off roots that seem to be kinked or broken.
• Dig the hole twice the width and depth of the root ball (Figure A-12B).

• Place some soil into the bottom of the hole, moisten, and tamp down.
• Spread roots out over the soil placed in the hole.
• Slowly fill in the hole with soil and water. The top of the root ball should be about two inches above surrounding soil (Figure A-12C).
• Make a soil berm to form a watering basin. Irrigate gently. Water should remain in the basin rather than flooding over the soil berm. It is important to keep the trunk base dry.
• Mulch around the plant base.
• Irrigate from the top, filling the basin with water and sprinkling around to settle soil and berm. Allow the water to soak in and repeat.

Maintenance Guidelines

Plan to water plants during the dry season for a minimum of three years. Some plants need to be dry in the summer while others may need a specific watering regime to survive. Discuss any of these maintenance requirements with the staff people of the nursery where you buy your plants. See Appendix N for information about protecting plants from animal damage.

[This appendix is reprinted with permission and some modifications from The Wetlands Conservancy, 1996.]

Figure A-12. (A) Container plants should be placed in (B) a large hole that (C) provides the root ball with ample room to gradually spread out and penetrate the surrounding soil matrix. A slight donut-shaped depression around the base of the newly planted shrub or tree allows water to collect and seep into the root zone. (The Wetlands Conservancy, 1996)

Appendix M

Bare-Root Plants

$B$are-root plants are plants that are grown in the ground, then dug out and sold to be transplanted somewhere else. In contrast to container planting, these plants are usually larger and the root systems have not been restricted by a container, usually making them healthier. Many of the deciduous plants, including fruit and shade trees and flowering shrubs, are available as bare-root stock.

Why Use Bare-Root Plants?

- Bare-root plants are usually larger and healthier than container plants because their root systems have not been restricted by a container.
- They tend to be less expensive than container plants.
- Bare-root shrubs and trees are planted earlier in the year than container plants. This gives them more time to become established and grow strong.

When and How to Get Plants

Bare-root plants can only be dug from December through April and they must be planted soon after being taken out of the ground. In winter and early spring, you can buy bare-root plants at many retail or mail-order nurseries. When buying plants, make sure that the roots are fresh and plump, not dry or withered.

How to Plant

- Gently unwrap the packaging around the plant roots, being careful not to damage the roots. Consult with the nursery on how to care for the roots. Soak roots overnight in water, or use a hose to moisten them.
- Spread roots out and place in the planting hole. Refill the hole with soil, burying roots just to the crown. A crown refers to the point at which a plant's roots and top structure join

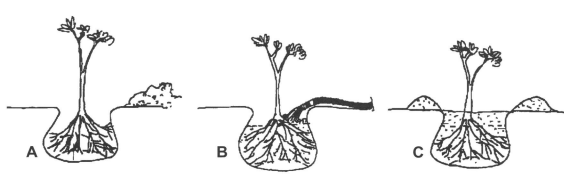

Figure A-13. (A) Dig a hole sufficiently large to hold the roots of the plant when they are spread out. When loosening the root mass of bare-root plants, take care to minimize root damage.(B) Make sure to water the plant and pack the soil before completely filling the hole. This ensures that any existing air pockets will be eliminated and that the roots will be firmly implanted in the ground. An advantage of using a bare-root plant is that its roots can establish themselves more quickly than the compressed roots of a container-grown plant. (C) A berm placed around the hole will facilitate the effective watering of the plant. (The Wetlands Conservancy, 1996)

(usually at or near the soil line) (figures A-13A, B).

- Fill in the hole with soil nearly to the top, firming it with your fingers as you fill, and then gently water. After the plant is placed correctly, fill in any remaining soil. Make a narrow berm of soil around the hole to form a watering basin (Figure A-13C).

Helpful Planting Tips

- Even if the roots appear fresh and plump, it is a good idea to soak the root system overnight in a bucket of water before you plant.
- Dig the planting hole broad and deep enough to accommodate the roots easily without cramping, bending, or cutting them to fit.
- Cut back any broken roots to healthy tissue.
- In areas with shallow or problem soils, a wider hole will help the plants take root faster.

Maintenance

After the initial watering, water bare-root plantings conservatively. Dormant plants need less water than actively growing ones, and if you keep the soil too wet, new feeder roots may not form.

Check the soil periodically for moisture and water as needed. If the soil around the root zone is damp, the plant does not need water.

When the weather turns warm and growth becomes active, water more frequently.

Do not over water. Check soils for moisture before watering. If hot, dry weather follows planting, shade the new plant at least until it begins to grow. And be patient; some bare root plants are slow to leaf out. Many will not do so until a few warm days break their dormancy.

[This appendix is reprinted with permission and some modifications from The Wetlands Conservancy, 1996.]

Appendix N

Protecting Plants from Wildlife Damage

A primary concern when maintaining riparian buffer plantings is the mortality or loss of growth caused by animal damage. An assessment of the animal activity in the planting area and the use of appropriate protection methods is the best insurance for plant survival. Deer and rodents cause most of the damage to riparian buffer plantings. The type of browse damage, droppings, and tracks can be used to make a positive species identification. After you determine the animals present at your stream enhancement site, you can choose control methods appropriate to the animal species involved.

The primary objective of any control program should be to reduce damage in a practical, humane, and environmentally acceptable manner. If you base control methods on the habits and biology of the animals causing damage, your efforts will be more effective and will serve to maximize safety to the environment, humans and other animals.

Deer

Deer can cause extensive damage to plantings. Deer enjoy the tender branches of young, newly planted trees and shrubs.

Although many stream enhancement planners choose to plant larger trees to prevent deer browse, these animals can reach up to 8 or 10 feet. In addition, bucks can girdle larger planted trees by rubbing off the bark with their antlers. Some project planners choose plant species that are less favorable for deer. However, the hungrier deer become, the less selective they are. Therefore, any plant is at risk to deer browsing (Figure A-14).

There are some effective methods for preventing or controlling damage caused by deer. One method is to install plastic tube guards over the plant to be protected (Figure A-15). Another effective and economical method for deterring deer is to surround a planting with a fence made of six-foot wooden stakes and fishing line.

Figure A-14. Browsing by deer is a threat to tender young tree branches as high as 8 to 10 feet from the ground. (Delaware Riverkeeper Network, 2003)

Drive four stakes into the ground to form a square around the plant. String 10- to 15-pound fishing line around the stakes at waist or chest height and pull very taut. The second line should be wrapped around the stakes at knee height (Figure A-16). This will protect larger trees whose leaves are above the browse line. For smaller trees whose leaves are within reach, use the same technique, but angle the stakes at 30 degrees and wrap the fishing line around the stakes in rows spaced approximately six inches apart. This will protect the leaves and shoots that are still within the browse area of the deer. The deer stop when they feel the line on their head, chest, and legs. However, they cannot see the line and do not realize it is a fence they can jump over. Remember to inspect the stake and line every few months or after a heavy storm and repair any damage.

Another method that can be effective in preventing deer browse is to spray the plants with a repellent. Many commercially-available repellents make plants taste bitter to deer.

Other repellents frighten deer by using blood or urine from predators. Repellents need to be re-applied after rain or other inclement weather. Unfortunately, most wildlife soon discover that repellents are not actually harmful, and the animals may soon become accustomed to the smell, taste, or feel of these deterrents.

In rural areas, hunting may also be an effective method to keep deer populations under control and therefore limit damage to young plants.

Rodents

Rodents, such as mice and rabbits, damage trees and shrubs by chewing the bark off of the trunks or nibbling on the

Figure A-15. Plastic tube guards prevent deer (and rodents) from destroying tree saplings and live stake cuttings. (Izaak Walton League of America photograph)

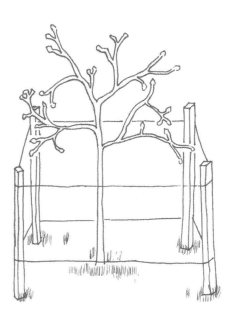

Figure A-16. Using wooden stakes and fishing line, this simple fencing technique can protect larger trees from deer damage. (Delaware Riverkeeper Network, 2003)

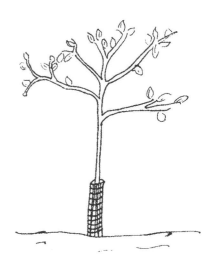

Figure A-17. Plastic or steel mesh guards are an effective defense against rodent damage. (Delaware Riverkeeper Network, 2003)

ends of branches. They can completely girdle a tree, causing it to die. Most of this damage is done during the winter when other food is not available.

Treeguards are an effective measure to protect trees and shrubs from rodent damage. The guards should be installed in the fall and removed in the spring, so they do not hinder plant growth. Plastic tube guards are the most widely used (Figure A-15). Other tree guard materials include plastic mesh and steel mesh (Figure A-17).

As with deer, rodent damage also can be discouraged with the use of commercially-available repellants. These can be applied to tree trunks with a paintbrush, and re-applied after rain or other inclement weather as necessary.

Appendix O

Watering

Plants require water to grow and survive. In the outdoors, plants receive water from rain and snow. However, for newly installed plants that are not yet established, additional watering in the hot and dry summer months is important.

Even native wetland and riparian plants often need supplemental watering. For the first three to five years, wetland plants need plenty of water to help them survive. After that, they can usually survive on seasonal moisture and begin shading each other to reduce evaporation.

Choose Planting Location Based on Water Needs

Plants adapt to their environment and seasonal watering if they are planted in the appropriate area. Some plants prefer wet or moist areas, while others like dry areas. Some plants like to be in the sun, while others prefer partial sun or shade.

In most planting projects, you will probably install different types of plants with different needs. Emergent plants such as rushes and sedges flourish in soil that is wet to moist year-round. Certain woody species, including alder and big leaf maple, need summer watering but prefer soil that dries out periodically. What you plant will determine the amount of water that is necessary and the time that it must be applied.

Emergent Plants

With emergents, you are generally using liner-grown plants (plants grown in small plastic tubes) or plugs (sections of the plant with roots dug out of the ground). Make sure you place them in an area of the site that will be wet to moist year-round. Water the newly installed plants, pack the soil back around the roots, and secure the plant in place.

Woody Plants

Whether you plant cuttings, container plants, or seeds, woody plants will require a good watering once they are installed

and additional watering during the late spring, summer, and fall months until they become established. By visiting the site many times after planting, you will determine when your plants need water.

Take Steps to Minimize Future Drought

Mulching around the base of newly planted plants helps hold water in the soil that thereby becomes available to the plant. Mulching also reduces weed growth.

Prepare a water basin with mounded soil in a tire shape around the plant to keep water near it during watering.

To plant understory and shade loving plants, consider waiting until the trees and shrubs are large enough to provide shade. This may take a few years, but it will eliminate competition for water during the critical first years of growth.

How to Water

Depending on your site size, budget, availability of volunteers, and plant needs, watering can be simple or extensive.

Hand Watering

It is important to create a basin around plants to ensure that water gets to the roots. Be sure to water the soil, not just leaves. This method is good if the site is small and near a water source.

Automatic or Manual Operating System

This watering method involves pipes underground that carry water to sprinkler heads throughout the site. The system can work automatically or manually. This method involves costs of system design, materials, and installation. It is the most efficient and reliable method, if your budget can justify it. Aboveground sprinkler systems can also be installed in your planting area and hooked up to a hose bib.

Drip Irrigation

This method involves a system of plastic tubes that are distributed throughout the site, with nozzles coming off the tubes that deliver drips of water to each plant. This method is less expensive than an automatic or manual operating system. Because the plastic tubes and nozzles are above ground, they are prone to vandalism or dirt getting into the nozzles, blocking the passage of water. Usually, the only time you know a nozzle is not delivering water to its plant is when the plant dies.

Flood Technique

If your site is a large, self-contained area, flooding it may be very effective. Before planting, prepare soil dikes on contour (following the lines of elevation) that will hold in the water. Once a month, flood the areas between the dikes with water and let it soak in. This method is very effective if it is suitable for the site.

[This appendix is reprinted with permission and some modifications from The Wetlands Conservancy, 1996.]

Appendix P

Control of Weeds

*T*o help new plants survive and become established, plan to remove or control weeds. You should remove weeds in the immediate area of planting, but you may also want to remove weeds upstream and downstream of your planting area, as weeds can spread seeds and reproduce rapidly.

What is a Weed?

A weed is any plant that is growing on the site that you do not want there. Weeds threaten the growth of new native plants. Some weeds also may be native plants. Others are exotic plants that have been planted by people or have been carried to the site by birds, wind, or water.

Why Remove Weeds From a Project Site?

These unwanted plants will compete with the newly installed plants. The weeds also compete with any native plants already there. Removing weeds and watering are the most important maintenance practices to make an enhancement project a success.

It is much easier to remove weeds before they get too large and become established. During the first two to three years, more weeds will need to be removed because of the larger areas of bare soil between the new plantings and the availability of weed seed sources in the soil. Also, the newly installed plants have not yet had a chance to grow and shade out the weeds.

It will be important to schedule weed removal days for the site according to how quickly the weeds are growing. Check for areas where weeds are starting to take over, especially around the newly planted trees and shrubs.

Methods of Controlling or Removing Weeds

Hand

Weeds can be controlled by pulling them out manually. Be sure to get all of the roots out of the ground, and remove the roots and all plant parts from the site. If you have many native plants growing among the weeds, hand control may be the most effective way to protect the native plants.

Mechanical

There are many tools that can assist you in removing weeds. These tools include lawn mowers, weed-whips, shovels, hoes, pick axes, and machetes. Using a tractor with soil attachments is another effective way to remove weeds from the site. Mechanical removal is most effective if the site is primarily weeds, or if the native plants are not close to the weeds. Tools and machines are helpful for removing plants with thorns that can be difficult to remove by hand.

Chemical

Chemical control is a very controversial topic. Be sure to understand the chemical and its impact on the immediate environment. If you choose to use chemicals, be sure to follow all the directions and precautions for safe application. If the weeds have become established and are very difficult to remove by other methods, consider using chemical control. Consult a member of your technical team when making this decision.

Solarization

This method is inexpensive and effective. Mow over the area of weeds and then cover with a sheet of dark plastic. This will stop the weeds from getting the sunlight they need to grow, and it creates a heat source that can kill the weeds.

Mulching

Mulching will assist in reducing weed growth around the newly installed plants, as well as retaining moisture. Mulch is typically a layer of straw or wood chips placed around the base of the plant.

[This appendix is reprinted with permission and some modifications from The Wetlands Conservancy, 1996.]

Appendix Q

Suggested Format and Contents for a Riparian Revegetation Plan

A final revegetation plan for riparian enhancement should contain the following information in order to ensure that the plan is well conceived, cost effective, and will result in the successful establishment of riparian vegetation.

Introduction

- Statement of goals and objectives.
- Location map(s) showing project site.
- Map or plan showing areas that will be revegetated.
- Discussions of any permit conditions or requirements affecting revegetation design and plan selection.
- Rationale for design (e.g., database used to select plant palette, plant composition, plating densities, pattern, etc.).

Analysis of Site Conditions

- Laboratory analysis and evaluation of soil samples.

- Description of soils and soil moisture.
- Discussion of soil problems and need for correction.
- Summary of available information on depth to groundwater and/or the results of field tests.
- Evaluation of rainfall data (as it may affect plant selection, choice of propagule, irrigation requirements, etc.).
- Anticipated extent and duration of flooding/inundation.
- Analysis of the effect of slope and aspect of areas to be planted on plant selection and planting plan.
- Determination of availability of water for irrigation.
- Evaluation of the effect of the above site conditions on plant selection, installation methods, etc.

Plant Materials

- List of plant species that will be installed (sometimes referred to as the plant palette).
- Selected or preferred type of planting stock to be utilized (e.g., seeds, cuttings, rooted cuttings, liners, tublings, one-gallon containers, etc.).
- Amount of each type of plant material required.
- Available/recommended commercial source(s) of plant materials.
- Localities available for collection of plant materials (seeds, cuttings, etc.).
- Lead time required for the procurement of plant materials.
- Recommended planting time.

Planting Design and Layout

(Sometimes called the Planting Plan)

- Recommended plant mixes/planting associations.
- Zones to be planted (with appropriate mixes).
- Desired percent composition for each plant species.
- Plant spacing and planting density.
- Design considerations (clustering of plants, etc.).
- Recommended seed mix and application rate for herbaceous ground cover (where appropriate for erosion control, weed control, habitat enhancement, etc.).

Site Preparation

- Grading and draining plans.
- Weed control.
- Removal of weeds.
- Pest control.
- Tillage.
- Soil augmentation.

Irrigation System

(Where appropriate)

- Type of irrigation (e.g., overhead sprinkler, flood irrigation, drip emitter, hand watering, etc.).
- System layout and specifications.
- Installation.

Installation of Plant Materials

- Purchase, transportation, and on-site storage of plant materials.
- Collection and handling of on-site material.
- Lay-out (flagging of planting locations).
- Methods of installation (e.g., water jet, augured holes).
- Schematic drawings of installation procedures.
- Application of fertilizer in planting holes.
- Watering basins, etc.
- Timing and coordination.

Plant Protection

(These should be implemented at the time of plant installation and/or during the establishment period.)

- Browse protection.
- Seed protection.
- Staking.
- Sun protection.
- Insect protection (e.g., grasshoppers).
- Weed control (e.g., fabric, mulch, etc.).

Maintenance during Establishment Period

- Length of establishment period.
- Control of weed competition.
- Acceptable level of mortality and required replacement of dead plants.
- Supplemental planting of additional areas and/or species not available in the first year.

- Frequency and amount of irrigation (including monitoring of necessity for irrigation).
- Other maintenance (e.g., fertilization, thinning of direct-seeded sites, etc.).
- Maintenance schedule.
- Potential for vandalism and strategy for control of vandalism.

Monitoring Procedure

(Where required)

- Documentation procedures during installation.

- Monitoring of plant survival, growth, and vigor.
- Monitoring of environmental conditions.
- Photo documentation of results.
- Periodic sampling of composition and cover.
- Monitoring of wildlife use.

Cost Analysis

- Labor and equipment needed for site preparation.
- Establishment period maintenance costs.
- Itemized budget.

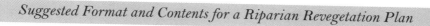

Suggested Format and Contents for a Riparian Revegetation Plan

Formalizing Your Group

*W*hile any group of friends, colleagues, or neighbors can work together to enhance local waterways, there are some advantages to formalizing the group. Formal groups can obtain tax-exempt status to help with fundraising. In addition, a formal group can gain credibility for its projects that can help when recruiting volunteers, establishing a technical team, working with the media, and obtaining permits.

One option for formalizing the group is to become a part of an existing group with a similar mission, such as a local watershed association or an Izaak Walton League chapter.

Another option is to incorporate in your state and become a 501(c)(3) tax-exempt organization. This means other people and organizations can give tax deductible donations to support your enhancement projects. Government agencies, private foundations, and other funders may only be able to give a grant to a not-for-profit organization. Tax-exempt status also allows the organization to make purchases without paying taxes on them.

Even if the group chooses not to apply for tax-exempt status, it is still a good idea to incorporate. A corporation is a legal entity that can exist separately from its owners. A nonprofit corporation is a corporation formed for purposes other than generating a profit and in which no part of the organization's income is distributed to its directors or officers. Corporations may be organized for religious, charitable, social, educational, recreational, or other purposes. Incorporation provides tax benefits and asset protection. Once the group becomes a corporation, group members are no longer personally responsible for the debts and liabilities of the corporation. Incorporation also provides structure for the group because it requires that you create bylaws and develop a board of directors.

Whether or not your group incorporates, there are many ways to create credibility within the community. One important way to look credible is to get a permanent mailing address. You may want to establish an e-mail address and web site for the group as well. The group may also create a letterhead, a logo, or a brochure. Most importantly, continue to work toward stream enhancement and improved water quality in your community.

For more information on incorporating and registering an organization for tax-exempt status, refer to the following resources:

- National Council of Nonprofit Organizations, *www.ncna.org.*
- "Establishing a Nonprofit Organization" on-line tutorial. This tutorial describes 12 tasks that need to be accomplished during the process of establishing a nonprofit organization. Issues include board development, creating bylaws, filing for federal tax exemption, recruiting staff, and developing an overall fundraising plan. The Foundation Center, *http://fdncenter.org/learn/classroom/establish/index.html.*
- "Tax Information for Charitable Organizations." The Internal Revenue Service, *www.irs.gov/charities/charitable/index.html.*
- The Internet Nonprofit Center, *www.nonprofits.org.* Includes a FAQ section for new organizations.
- The Alliance for Nonprofit Management, *www.allianceonline.org.*

Appendix S

Liability Considerations

Volunteers are valuable assets to any organization. Therefore, an organization's leadership needs to consider the safety and health of its volunteers when planning projects and events. Planning for safe events is an important step in preventing volunteer injuries. However, organizations also need to protect themselves and their volunteers when injuries do occur. Although many people think charitable organizations are protected from lawsuits, they are not. About 40 years ago, nonprofits were protected from liability under a doctrine called "charitable immunity." However, courts have gradually abolished the practice of charitable immunity, and nonprofits now are required to assume legal and financial responsibility for their activities. Under tort law, an injured volunteer may recover monetary damages from the person or organization that caused or contributed to the harm. Damages are based on the extent of loss,

including pain and suffering. Although lawsuits are not common, when they do occur they can be devastating to an organization's finances and credibility.

The first step in protecting volunteers and the organizations they serve is to prevent injuries by planning for safety at events. See Appendix D for additional information. Inform volunteers about safety concerns and procedures. Also, inform them about any coverage or other provision the organization has made for handling injuries.

It is a good idea to have participants sign a waiver or an informed consent agreement before participating in the activity. Liability waivers are written agreements in which the volunteer is informed of the specific risks involved in the activity and waives the right to sue an organization. Liability waivers are legally vulnerable because these agreements are only valid if the person enters the agreement knowingly and voluntarily. Often, an agreement between an organization and an individual is not considered voluntary because the parties are of unequal bargaining power. However, if properly drafted and executed, waivers may help block liability. In addition, an individual who has signed a waiver may be less likely to initiate a lawsuit than someone who has not. Remember that people under 18 do not have the capacity to sign contracts. Instead, have a parent or legal guardian sign a permission slip.

Obtaining Insurance Coverage

First check any existing insurance coverage provided by your organization to see if volunteers are covered. If volunteers are not covered under an existing policy, discuss options with your insurance company for adding them to the policy. Look for insurance packages that include:

- Public liability. This covers the organization, its staff, and volunteers for the legal liability to third parties for personal injury or damage to personal property.

- Personal accident insurance. This covers volunteers for costs related to an injury that occurs while volunteering. Check policies carefully because some have age limits or other considerations.

- Professional indemnity insurance. This covers the organization when staff or volunteers have given negligent advice or service leading to injury. If your organization currently does not have an insurance policy, or if you are interested in exploring other options for volunteer coverage, consider the following options:

- Purchase a special event policy. A one-day special event policy typically provides injury compensation coverage of $2,000–$3,000 per person per accident at a premium of approximately $60–70. This coverage does not provide tort liability for the sponsoring organization, which would be much more expensive. This may be considered more of a "good will" policy. When using this type of policy, ask volunteers to check the coverage offered for injury compensation under their family major medical policy before participating in the activity.

- Purchase a general liability policy. Investigate several companies to compare coverage and costs. Some underwriters will cover special events, like river cleanups, in their general policy while others do not. If your general liability policy will not cover all special events, one option is to get a temporary or permanent special event rider on your existing policy. Search for insurance companies in the local phone book. Some insurance companies specialize in providing volunteer coverage.

For example, an insurance company called CIMA offers volunteer insurance including up to $25,000 for medical expenses at a cost of $3.75 per volunteer per year. Also available is up to $1,000,000 in personal liability insurance for $1.50 per volunteer per year, and up to $500,000 in excess automobile liability beyond the volunteer's own insurance for $5.25 per volunteer per year. For additional information, visit *www.cimaworld.com.*

- Join the Natural Resources Conservation Service Earth Team. This volunteer program extends liability protection to

event sponsors and injury protection to volunteers under the agency's workers compensation package. Volunteers must be 14 years of age to qualify. The Natural Resources Conservation Service requires a group application to be filed in advance and a record of all participants to be maintained on the day of the event. Contact your county Natural Resources Conservation Service office or the Natural Resources Conservation Service Earth Team coordinator for your state. Additional information is available at *www.nrcs.usda.gov/feature/volunteers*.

Additional Resources on Managing Risk

• Internet Nonprofit Center. This web site has an excellent FAQ section with information on the organization, management, regulation, resources, and development of nonprofit organizations, including a section on liability. Visit *www.nonprofits.org*.

• Nonprofit Risk Center. This center offers a free download of a report called "State Liability Laws for Charitable Organizations and Volunteers." Visit *www.nonprofitrisk.org*.

• On-Line Nonprofit Organization and Management Development Program. This free on-line course includes several modules on risk management and liability coverage for volunteers. Visit *www.mapnp.org/library/ np_progs/org_dev.htm*.

• Risk Management Resource Center. This group helps local governments, nonprofit organizations, and small businesses manage risks effectively by making risk management knowledge and practical information available on-line. Visit *www.eriskcenter.org*.

Bibliography

Ammann, A., and A. L. Stone. 1991. *Method for the Comparative Evaluation of Nontidal Wetlands in New Hampshire.* Concord, NH: New Hampshire Department of Environmental Services.

Boettcher, Susan, and others (compilers). 1998. *The Mortenson Ranch: Cattle and Trees at Home on the Range.* Brookings, SD: South Dakota State University College of Agriculture and Biological Sciences.

Bolling, David. 1994. *How to Save a River: A Handbook for Citizen Action.* Washington, DC: Island Press.

Buckhouse, J. 1995. "'Off-site' livestock watering alternatives." *Western Beef Producer,* March 2.

Chesapeake Bay Program. 1998. *Habitat Requirements for Chesapeake Bay Living Resources: A Report from the Chesapeake Bay Executive Council.* Annapolis, MD: Chesapeake Bay Program.

DeBerry, D., ed. 2001. *Stream Steward Restoration Guide: A Small Woodland Owners Guide to Stream Habitat Restoration.* Washington, DC: American Tree Farm System and Trout Unlimited. *www.treefarmsystem.org.*

Delaware Riverkeeper Network. 2003. *Adopt-a-Buffer Toolkit: Monitoring and Maintaining Restoration Projects.* Washington Crossing, PA: Delaware Riverkeeper Network. *w.delawareriverkeeper.org.*

Dunne, T., and L. B. Leopold. 1978. *Water in Environmental Planning.* San Francisco, CA: W. H. Freeman and Company.

Federal Interagency Stream Restoration Working Group. 1998. *Stream Corridor Restoration: Principles, Processes, and Practices.* Washington, DC: Government Printing Office. *http://www.usda.gov/stream_restoration/.*

Firehock, K., and J. Doherty. 1995. *A Citizen's Streambank Restoration Handbook.* Gaithersburg, MD: Izaak Walton League of America.

Firehock, K., L. Graff, J. V. Middleton, K. D. Starinchak, and C. Williams. 1998. *Handbook for Wetlands Conservation and Sustainability.* Gaithersburg, MD: Izaak Walton League of America.

Fitch, L., B. W. Adams, and G. Hale, eds. 2001. *Riparian Health Assessment for Streams and Small Rivers — Field Workbook*. Lethbridge, Alberta: Cows and Fish Program.

Hanley, D., and Kuhn, G. 2003. *Trees against the Wind*. Pullman, WA: Pacific Northwest Cooperative Extension Publication PNW5.

Harrelson, C. C., C. L. Rawlins, and J. P. Potyondy. 1994. *Stream Channel Reference Sites: An Illustrated Guide to Field Technique*. Fort Collins, CO: USDA Forest Service, Rocky Mountain Forest and Range Experiment Station.

Izaak Walton League of America. 2004. "Watershed Stewardship Action Kit." Gaithersburg, MD: Izaak Walton League of America.

Marcy, L. E. 1986. *Special Wire Fences: US Army Corps of Engineers Wildlife Management Manual*. Vicksburg, MS: US Army Corps of Engineers Waterways Experiment Station.

Montgomery, D. R., and J. M. Buffington. 1993. *Channel Classification, Prediction of Channel Response, and Assessment of Channel Conditions*. Washington State Department of Natural Resources, Timber/Fish/Wildlife Agreement, Rpt. TFW-SH10-93-002.

Pfankuch, D. J. 1975. *Stream Reach Inventory and Channel Stability Evaluation*. Missoula, MT: USDA Forest Service.

Rosgen, D. 1996. *Applied River Morphology*. Pagosa Springs, CO: Wildland Hydrology.

Schumm, S. A. 1977. *The Fluvial System*. New York, NY: John Wiley and Sons.

Smith, K., D. O'Connor, and D. Uchiyama. 1995. *Tualatin River Basin Stream Enhancement Handbook*. Hillsboro, OR: Clean Water Services.

The Wetlands Conservancy. 1996. *Wetland Restoration: Steps to Success*. Tualatin, OR: The Wetlands Conservancy. (Video)

US Department of Agriculture, Forest Service. 1995. *Guide to the Identification of Bankfull Stage in the Western United States*. Fort Collins, CO: USFS Stream Systems Technology Center, Rocky Mountain Forest and Range Experiment Station. (DVD and VHS)

US Department of Agriculture, Forest Service. 2003. *Identifying Bankfull Stage in Forested Streams in the Eastern United States*. Fort Collins, CO: USFS Stream Systems Technology Center, Rocky Mountain Forest and Range Experiment Station. (DVD and VHS)

US Environmental Protection Agency. 1996. *The Volunteer Monitor's Guide to Quality Assurance Project Plans*. Washington, DC: US Environmental Protection Agency.

US Environmental Protection Agency. 1994. *The Water Quality Standards Handbook*, Second Edition. Washington, DC: US Environmental Protection Agency.

West, J. 1995. *Restoring the Range: A Guide to Restoring, Protecting and Managing Grazed Riparian Areas*. Gaithersburg, MD: Izaak Walton League of America.